AF602046

Sustainable Agriculture and Natural Resource Management

Issues and Challenges

THE EDITORS

Dr P. K. Mishra is currently Associate Professor and Head of the Department of Economic Studies, Central University of Punjab, Bathinda. He has about 20 years of teaching and 10 years of research experience. He has to his credit 11 books, 4 edited volumes and more than 110 research papers published in journals of national and international repute. His areas of specialization include quantitative economics, development economics, financial economics, tourism economics, and resource economics. He is the life member of The Indian Econometric Society and Orissa Economic Association. He is the acknowledged review board members of a number of prestigious national and international journals. He has delivered lectures and presented research papers at several national and international conferences.

Dr J. K. Verma is currently Assistant Professor in the Department of Economic Studies, Central University of Punjab, Bathinda. He has about five years of teaching and research experience. He has to his credit 4 edited volumes and about 20 research paper publications in national and international journals of repute. His areas of specialization include development economics and entrepreneurship. He is the life member of Indian Economic Association, Indian Commerce Association, The Indian Econometric Society, Indian Association for the Study of Population, Agricultural Economics Research Association and Uttar Pradesh, Uttarakhand Economic Association. He is associated with a number of Journals as editorial board members and reviewers. He has presented papers in a number of national and international conferences.

Sustainable Agriculture and Natural Resource Management

Issues and Challenges

P. K. Mishra

J. K. Verma

2019

Daya Publishing House®

A Division of

Astral International Pvt. Ltd.

New Delhi – 110 002

ISBN: 9789389569148 (Int. Edition)

Published by : **Daya Publishing House®**
A Division of
Astral International Pvt. Ltd.
– ISO 9001:2015 Certified Company –
4736/23, Ansari Road, Darya Ganj
New Delhi-110 002
Ph. 011-43549197, 23278134
E-mail: info@astralint.com
Website: www.astralint.com

Digitally Printed at : **Replika Press Pvt. Ltd.**

Preface

Right from the ancient time, the concept of sustainability has been in the centre stage of socio-economic and environmental development of a culture-rich nation like India. The typical Indian work culture, lifestyle, food & beverages, fairs & festivals and in everything else respect for the environment and a consciousness about the rational use of resources are often noticed. However, along with the structural transformations in the social characteristics, and with the rapid growth of population and the economy, the demand for habitations, household articles, food etc. have been multiplied at a compounded rate. It has necessitated the economizing of the use of available resources and the environment while amicably catering to the needs of the present and future generations. This concept is all about typical management practices that are applicable to business, agriculture, society, environment and personal life such that both the present and future generations are benefitted; hence, appropriately called the sustainable management practices. The primary aim of the concept of sustainable management is to explore and adopt the viable practices and activities that fit into the carrying capacity of the Earth.

In this context, this volume brings together scholarly perspectives on sustainable agricultural practices and natural resource management and the associated problems and prospects. This book will certainly be instrumental for all those interested in understanding the contemporary issues and challenges associated with the agricultural sustainable management practices.

In this connection, we take the opportunity to express our high gratitude to the authors of the chapters who have been meticulous in imparting a distinct scholarly quality to this publication.

We sincerely express our gratefulness to our patron, Prof. R. K. Kohli, Hon'ble Vice-Chancellor, Central University of Punjab, Bathinda who has been a source of unceasing inspiration and unparalleled encouragement in our entire academic and research initiatives.

We extend the deepest gratitude to Prof. P. Ramarao, Dean, Academic Affairs, Central University of Punjab, Bathinda who has constantly tendered valuable advice on all key matters concerning academics and research.

We are thankful to Dr. Jagdeep Singh, Registrar, Central University of Punjab, Bathinda and Mr. K.P. S. Mundra, Finance Officer, Central University of Punjab, Bathinda for their all-round support in carrying out our academic and research events, activities and responsibilities.

We also take this opportunity to owe our heartfelt thankfulness to all of our colleagues who have been the constant sources of motivation, encouragement and support in bringing this volume to the limelight.

We sincerely acknowledge the support of Indian Council of Social Science Research (ICSSR), New Delhi for approving financial assistance to the Central University of Punjab, Bathinda to deliberate on various aspects of sustainable management practices. We earnestly express our thanks for the gracious patronage of ICSSR.

Last but not the least, we express our thanks to the entire editorial team of Astral International Pvt. Ltd., New Delhi for their meticulous and untiring cooperation and endeavour for timely publication of this edited volume.

P. K. Mishra

J. K. Verma

Contents

List of Contributors

Amrita Pani
Utkal University, Bhubaneswar, Odisha

Ashish Jha
Rajiv Gandhi National Institute of Youth Development, Sriperumbudur, Tamil Nadu

Sonali Sharma
Shri Mata Vaishno Devi University, Katra, Jammu & Kashmir

Kakali Majumdar
Shri Mata Vaishno Devi University, Katra, Jammu & Kashmir

Navneet Kaur Nijjar
Central University of Gujarat, Gandhinagar, Gujarat

C. Prakasam
Chitkara University, Himachal Pradesh

R. Saravanan
Chitkara University, Himachal Pradesh

Vijayachandra Reddy S.
Agriculture College, Kalaburagi, Karnataka

S. M. Mundimani
Agriculture College, Bijapur, Karnataka

Harpreet Kaur
Mata Sahib Kaur Girls College, Talwandi Sabo, Punjab

Harminder Singh
Central University of Punjab, Bathinda, Punjab

L. T. Sasang Guite
Central University of Punjab, Bathinda, Punjab

Dipika Basu
West Bengal State University, Kolkata, West Bengal

Arun Kumar Nandi
Chakdaha College, University of Kalyani, West Bengal

Sulata Hembrem
Basirhat College, West Bengal State University, West Bengal

Partha Pratim Roy
Mahatma Gandhi College, Sidho Kanho Birsha University, West Bengal

Uttam Haldar
West Bengal State University, Kolkata, West Bengal

Binali
Central University of Punjab, Bathinda, Punjab

Chidanand Patil
Central University of Punjab, Bathinda, Punjab

Chaitra G. B.
University of Agricultural Sciences, Dharwad, Karnataka

S. N. Tripathy
Gokhale Institute of Politics and Economics, Pune

Arshdeep Kaur
Central University of Punjab, Bathinda, Punjab

Ravindra Singh
Central University of Punjab, Bathinda, Punjab

Gaurav Kumar
Central University of Punjab, Bathinda, Punjab

Amandeep Kaur
Central University of Punjab, Bathinda, Punjab

Kiran Kumari Singh
Central University of Punjab, Bathinda, Punjab

Amit Kashyap
Central University of Punjab, Bathinda, Punjab

Shashi Bala Kashyap
S. C. D. Government College, Ludhiana, Punjab

Syed Wajid-Ul-Zafar
Central University of Punjab, Bathinda, Punjab

1

Sustainable Agriculture and Natural Resource Management

P. K. Mishra and J. K. Verma

Introduction

In the present era of globalization and information technology revolution, a large pie of the world population still lives in rural areas amidst small holdings of lower infertile land, and thus suffers from several socio-economic problems including undernourishment, unemployment, inequality and poverty. Consequently, they are often deprived of decent social participation and enjoying a better quality of life. Even today, agriculture is the major source of food supply within the world supply chain. But due to the growing population along with the limited availability of land and water, the world economy is facing the threats of food insecurity. Besides, the presence of elevated levels of CO_2 and other GHGs in the environment, and the consequential effects of climate change in terms of unpredictable weather extremes are all adding feathers to the problems of food insecurity. In view of these problems, when people attempt to maximize agricultural production by intensifying the land use and applying chemical fertilizers and pesticides, the soil fertility is further reduced and it becomes vulnerable to erosion due to poor soil management practices causing damages to healthy ecosystems thereby limiting the future agricultural productivity.

Researchers, academicians and policy-makers across the globe suggest for implementing sustainable agricultural practices and natural resource management to get rid of the problems of food insecurity and to escape from poverty traps. Sustainable agriculture and natural resource management are those activities which look for increasing agricultural productivity through the adoption of practices that maintain the long-term ecological and biological integrity of natural resources.

The objective here is the efficient utilization of land and other natural resources including water, air, minerals, forests, etc. for providing better economic returns and contributing to the improved quality of life and economic development.

Thus, sustainable agriculture and natural resource management system aims to address the issues concerning adapting to climate change, ensuring increased food supply, reversing environmental degradation, and promoting sustainable practices in the use of land and other natural resources. Precisely, sustainable agricultural practices aim to bring positive outcomes in terms of increased food productivity, reduced pesticide use and balanced CO_2 in the atmosphere.

Sustainable agriculture can be thought of as an agricultural or farming system which is capable of maintaining the productivity and usefulness of the land and other natural resources to society indefinitely. Such a system is required to be resource conserving, socially supportive, commercially competitive, economically lucrative and environmentally feasible. Thus, sustainable agriculture aims to ensure simultaneously the economic, environmental and social sustainability. Economic sustainability is ensured when the dependence on automation and chemical fertilizers and pesticides is minimized. Environmental sustainability is achieved when people use natural fertilizers and pesticides, minimize water usage, reduce soil erosion, and manage the cropping pattern in line with the changing weather conditions. Finally, social sustainability is achieved when sustainable agriculture and natural resource management system enhances the use of manpower thereby solving the problems of undernourishment, unemployment, inequality and poverty. Therefore, it is apt to say that the primary concerns of sustainable agricultural systems are to develop technologies and practices that are environment-friendly, accessible to and cost-effective for farmers, and reduce the vulnerability to food insecurity.

It is with this backdrop, this edited volume compiles the scholarly writings of experts in the field in 14 chapters which evaluates the sustainable agricultural practices and natural resource management system focusing on the underlying issues and challenges.

Overview

The lead paper *Sustainable Management Practices in Agriculture: A Case of Organic Farming in India* by Amrita Pani evaluates the issues and challenges pertaining to the growth of organic farming as an important sustainable management practice in Indian agriculture.

Ashish Jha in his paper *Sustainable Food and Livelihood Security: A Case of an Agricultural Intervention in Bangladesh* evaluated a unique sustainable agricultural practice adopted in the flood-prone areas of Bangladesh which has brought significant changes in the lives of marginal farmers and/or landless families.

Sonali Sharma and Kakali Majumdar in their paper *Productivity and Efficiency of Asian Agriculture* evaluate the productivity and efficiency of Asian agricultural countries in order to see whether agricultural productivity in Asia is sustainable.

Navneet Kaur Nijjar in her paper *Climate Change and Sustainable Food Security* analyses the agricultural practices causing climate change and further develop solutions within these practices so as to mitigate the ill-effects of climate change.

C. Prakasam and R. Saravanan in their paper *Efficiency of Agricultural Practices in Una District, Himachal Pradesh* evaluate the existing rural practice in the Una locale and coping up strategy pursued by the ranchers to sustain the farming.

Vijayachandra Reddy S. and S. M. Mundimani in their paper *Resource Use Efficiency and Comparative Costs: A Case of Farming System in Karnataka* examine the resource use efficiency and the comparative costs of organic farming systems in the Bijapur district of Karnataka state as a means of addressing the problems of food insecurity.

Harpreet Kaur in her paper *Sustainable Agriculture Development in Punjab* discusses the strategies adopted in the Punjab state of India for achieving sustainability in agriculture.

Harminder Singh and L. T. Sasang Guite in their paper *Prospect of Sustainable Agriculture in Punjab* evaluate the present agricultural scenario in the Punjab state and by analyzing the geo-physical factors focus on the issues pertaining to food security.

Dipika Basu, Arun Kumar Nandi, Sulata Hembrem, Pratha Pratim Roy and Uttam Haldar in their paper *Role of Education in Sustainable Agricultural Marketing Practices in Indian Horticulture* examine the contribution of education and extension services to sustainable agricultural practices in Indian horticulture.

Binali, Chidanand Patil and Chaitra G. B. in their paper *Sustainability of Agri-tourism in Punjab: Problems and Prospects* analyze agri-tourism as a strategy of achieving agricultural sustainability in the Punjab state of India.

S. N. Tripathy in his paper *Natural Resource Management for Environmental Sustainability* discusses the issues pertaining to the efficient management of natural resources for achieving environmental sustainability.

Arshdeep Kaur and Ravindra Singh in their paper *Remote Sensing and GIS in Natural Resource Management: A Case of Forest Cover in Punjab* focus on the efficient management of natural resources using remote sensing and GIS techniques and mapping of forest cover in the Punjab state for ensuring sustainable development.

Gaurav Kumar, Amandeep Kaur and Kiran K. Singh in their paper *Sustainable Development Goals and Wetland Sustainability: A Case of Harike Wetland in Punjab* examine the wetland sustainability and multidimensional significance of the wetland for the ecosystem.

Amit Kashyap, Shashi Bala Kashyap and Syed Wajid-Ul-Zafar in their paper *Sustainable Development: A Harmonious Link between the Pillars of Environment and Development* highlighted that the sustainable development of a nation is contingent upon the harmonious link between the environmental and socio-economic factors.

2

Sustainable Management Practices in Agriculture: A Case of Organic Farming in India

Amrita Pani

Introduction

"The discovery of agriculture was the first big step towards a civilized life"

– Arthur Keith

India is an agrarian country where agriculture including both cropping and live stock farming is the prime source of livelihood for about 58 % of population and considered as a noblest profession as well. Agricultural sector plays a major role for growth and development of the country through contribution of 17-18 percent GDP with the growth rate of 2.1 per cent according to the Indian Economic Survey 2017-18. During Financial Year 2010 to 2018 the total cost of agricultural exports from India increased at a CAGR of 16.45 % and ultimately reached to US$ 38.21 billion in 2018. In the present global competitive scenario, the trend of agricultural business has undergone a paradigm shift due to revolution in green technology. However, contemporary agricultural practices also encounter with sustainability issues due to excessive use of chemical ingredients in farming procedure. Mostly in synthetic-based farming, the utilization of chemical inputs have resulted in several unsustainable activities like soil erosion, environmental pollution, water contamination, genetic erosion and health disorders which further leads to financial loss to the farmers (Reddy, 2010).

Accordingly, agricultural sustainability is highly affected if there is a slight disproportion in the ecosystem and biodiversity functions. Therefore, to address the sustainability challenges, the idea of organic farming came in to picture and widely accepted throughout the globe. The modern agriculture revolution has incorporated sustainability practices to ensure healthy earth, healthy food and healthy populace. Sustainable agriculture is basically a pro-nature, pro-poor and pro-women approach that consists of an amalgamation of eco-friendly technologies, traditional wisdom, farmer's cooperation, policy makers, government, market linkage etc. (Keshavan and Swaminathan, 2008).

In other words, sustainable agriculture is all about balancing between People, Planet and Profit (Vyavahare and Bendal, 2012).

Similarly, with the increasing craze for unadulterated agro-products gradually the concept of organic philosophy also initiated to be deployed in the mainstream of agriculture to nurture growth sustainability and global competitiveness. The concept of organic agriculture was originated with belief system of healthy farm will give rise to healthy inhabitants. In general, organic farming notion is a holistic approach aims to produce qualitative food products as well as balancing the natural interlinked bionetwork. Organic farming depends upon stabilizing agro-ecosystems, maintaining ecological balances, developing biological processes to their optimum, and linking agricultural activities with the conservation of biodiversity (Alfoldi et al. 1992). Thus, organic farming is the environmentally sustainable agricultural system which balances the biodiversity, restricts exploitation of renewable resources, conserving soil health and fertility and also creates sustainable lifestyles for the forthcoming generation (Sowmya, 2014). In a nutshell, organic farming is all about a holistic system to cultivate a healthy agro-ecosystem, enhancing biodiversity and soil fertility (Dubey and Shukla, 2014).

In present context, organic farming is one of the most discussed issues of the century by providing an alternative solution of farming through unveiling the cons of non-organic chemical farming (Nemes, 2009). Research has also proved that, organic farming has more holistic value as compared to conventional chemical farming. Organic harvesting has many potential benefits ranging from environment friendly to reduction of soil erosion and nurturing biodiversity. Due to utilization of natural harvesting procedure with existing and recyclable resources; it is also cost effective for the farmers and agripreneurs. Organic cultivation further helps in enhancing the fertility of soil through the application of green manure and vermiculture using bacteria, insects and earthworms. Food items manufactured through chemical farming also having adverse effects to the health of the consumer due to use of hazardous synthetic pesticides, fertilizers and hormones. The contaminated land and water negatively influence livestock that eventually deteriorate the health of human being after consuming their by-products. Hence, medical practitioner now a days recommends the patients to go for organic consumption to avoid carcinogenic effects of pesticide contamination. Besides, different innovative methodologies and certification procedures also helps the farmers to enhance their self confidence and economic independence.

This research study focuses on the significance of organic farming in sustainable agricultural practices in India. The study analyses the role of organic farming to accomplish agricultural sustainability and inclusive growth through the application of innovative indigenous techniques. The scope of the study is limited to organic farming in India only.

Objectives and Methodology

The objectives of this paper are: (i) to study the scenario of sustainable management practices in agriculture through organic farming in India, and (ii) to analyze various scopes and challenges in the field of organic farming sector in Indian environment.

This research is a conceptual study based on application of secondary data analysis. The data has been collected from government reports, published statistics, journals, earlier research and internet.

Sustainable Agricultural Practices and Organic Farming

In this ferocious competitive economy strategizing innovation through sustainability approach is the key mantra for achieving global competitiveness and growth inclusivity for each and every organization. Through the application of principles of sustainability and effective utilization of local resources, organic agriculture has results in to the socio-economic development of the nation (Kilcher, 2007). Therefore, in agricultural business innovative sustainability practices have been incorporated in the business policy, to accomplish competitive advantage without disturbing the natural ecosystem. There are many smart techniques are deploying in agricultural field to increase sustainability through environmental management under the banner of intelligent infrastructure. The increasing awareness for organic products has given rise to the emergence of organic agriculture in the global forum. Organic farming has four basic dimensions i.e. health, ecology, fairness and care. Organic farming system through eco-friendly techniques not only helps in soil productivity through eliminating the harmful effects of chemicals but also propels recycling of agricultural waste and residues to enhance bio diversity.

In India, the concept of organic farming was initiated since the ancient Vedic age following the doctrines of plant science, *Vrikshayurveda*. Ancient version of organic farming or *Vedickrishi* is based on the principles of Cowpathy, is an agricultural process without using any harmful chemicals and pesticides to elevate yield of the crop (Naresh et al. 2018). It mostly uses cow's excreta as a base, and then includes other natural ingredients like dried neem or Margosa leaves, turmeric powder, honey, raw milk, curd, etc. and further utilize them for seed and soil treatment as well as pest repellent. Some of the significant procedures of *Vedickrishi* includes, *Panchagavya* (mixture of cow dung, urine, milk, curd and clarified butter/ghee), Vermiwash (mixture of cow dung and earthworm), *Beejamrit* (mixture of cow dung, urine, milk, lime stone and water), *Jeewamrit* (mixture of cow dung, urine, jaggary, pulse flour, organic soil and water), *Amritpani* (mixture of cow dung, butter, honey, water). With applications of these above non chemical green fertilizers in organic farming significantly improves the horticultural scenario. Likewise, in vedic farming model herbal extracts from neem, chilli, turmeric, asafoetida, basil garlic etc. effectively used as bio-pesticide foliar sprays namely *Bramhastra, Agneyastra, Neemastra*. Interestingly, in many a case *Vedickrishi* through *Homa* farming also known as *Agnihotra* based upon performance of *Yagna* or *havan* to enhance soil fertility, seed health and disease resistance in a drastic manner.

Organic farming, as defined by International Federation of Organic Agriculture Movement (IFOAM), "is the production system that sustains the health of soils, eco-systems and people. It relies on ecological processes, bio-diversity and cycles adapted to local conditions, rather than the use of inputs with adverse effects." Moreover, organic farming is a combination of tradition, innovation and science to benefit the environment and improve quality of life. According to The Food and Agriculture Organization (FAO), organic farming can be defined as "a holistic production management system which promotes and enhances agro-eco-system health, including biodiversity, biological cycles, and soil biological activity using, agronomic, biological and mechanical methods, as opposed to using synthetic materials". Similarly, Ministry of Agriculture, defined organic farming "as a system of farm design and management to create eco-system, which can achieve sustainable productivity without the use of artificial external inputs such as chemicals, fertilizers and pesticides." Niggi (2015) declared organic farming as the concept which combines agro-ecological approaches with productivity.

Hence, organic cultivation is the method of utilizing different innovative techniques for preservation of soil fertility and nutrient cycle through eco-friendly ingredients.

However, the method of organic farming varies from climate to climate and region to region. But the common objective of the system is to cultivate in a healthy and greener way without interfering with the environmental protocol. Accordingly, it has also been scientifically verified that one of the most sustainable way of agriculture is organic farming in form of protecting the eco-system.

Due to rising demand for nutritious and healthy food products, the scope for organic farming is also increasing. Many business giants are now showing interest to invest in this field.

International Federation for Organic Agriculture Movements (IFOAM) also plays a pivotal role to promote agribusiness of organic products in global context. The global organic food market is growing at a remarkable pace. As reported by Organics International and the Research Institute of Organic Agriculture (FiBL & IFOAM), 2018; global organic market reached to approximately 90 billion US dollars in 2016. From the survey including 178 countries, it was found that around 57.8 million hectares land have been certified as organic with a growth rate of 7.5 million hectares. As depicted in Figure 1, Australia has the majority in organic cultural lands i.e. 27.15 million ha. In global context, an European country like Liechtenstein has the highest percentage of organic share (37.7 percent) of the total farmland .Though, countries like Australia, Latin America, Asia are significant producers of organic crops they have not captured a large market share. Whereas, US has been considered as the market leader for organic products (38.9 billion euros). On the other hand, Switzerland topped the list in case of highest per capita of organic food consumption (274 Euros and highest shares the organic market of the total market is in Denmark (9.7%).

Fig. 2.1: Top Ten Countries with Organic Cultural Lands in 2016

Source: *FiBL and IFOAM Report, 2018, Retrieved from http://orgprints.org/32677/19/Willer-2018-global-data-biofach.pdf*

Organic Farming Scenario in India

In India, the concept of organic farming emerged with the increasing demand for organic food products inspired from western countries. Gradually, many modern entrepreneurs took interest in the sustainable agribusiness movement to provide holistic food products without harming the bio diversity. Furthermore, the consumer purchasing preference also inclines towards the products with organic labels. India also has a significant level of potentiality for the growth of organic cultivation due its highest number of agro climatic regions.

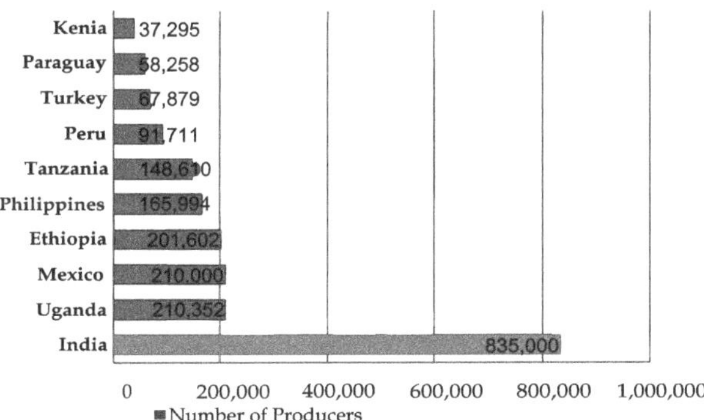

Fig. 2.2: Top Ten countries with the Largest Numbers of Organic Producers as on 2016

Source: *FiBL and IFOAM Report, 2018, Retrieved from http://orgprints.org/32677/19/Willer-2018-global-data-biofach.pdf*

According to FiBl and International Federation for Organic Agriculture Movements (IFOAM) Survey 2018, the total organic agricultural land in Asia was 4.9 million ha in 2016; while, more than 30 per cent of total number of organic producers are from India only. Similarly, according to World of Organic Agriculture Report 2018, from 2.7 million organic producers from all over the world, 46 % of producers were from Asia among which India has the major producers (835,000) which is clearly seen from Figure 2 and 3. Moreover, India constituted 3.56 million ha of certified land including 1.78

million ha of cultivable land and 1.78 million ha of non cultivable wild forest area as on 2017-2018. Furthermore, Madhya Pradesh has highest ranking due to having largest number of organic certified land (4.62 lakh ha).

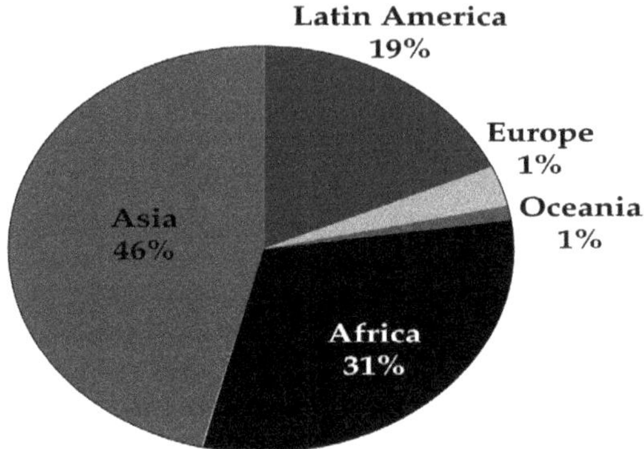

Fig. 2.3: Organic Producers by Region, 2016

Source: *Retrived from https://www.slideshare.net/francoisstepman/the-contribution-of-organic-agriculture- in - the - tropics - to - sustainable - development , Accessed on 5th September, 2018.*

Similarly, among Asian countries India is the second largest exporter of organic products succeeding to China. According to the annual report of APEDA, during 2017-18 a total volume of 4.58 lakh MT of organic food products were exported from India at the price of 515.44 million US Dollars (Figure 4). The highest category of organic food eligible for export comprises oilseeds (1.32 lakh tons exported) followed by Cereals and millets (0.44 lakh tons) and Processed foods (0.67 lakh tons). A sum total of 1.70 million MT of certified organic products from different categories were manufactured during 2017-18 in India.

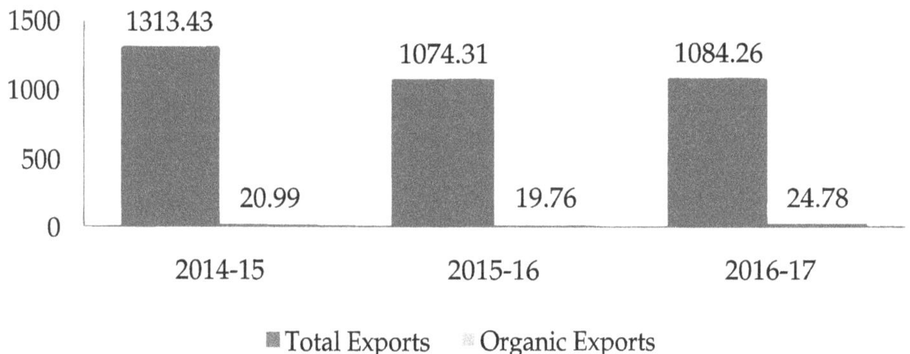

Fig. 2.4: Overall Farm Exports and Organic Exports (Rs. Billion)

Source: ***Retrieved from https://www.business-standard.com/article/economy-policy/organic- food - exports-surge-certification-remains-a-major-issue-118032800261_1.html***

Figure 5, depicts some of the selected category of organic crop production in Tons and their export quantity in Lakh MT in 2016-17. The highest category of organic crops produced in 2016-17 were belongs to (Oil seeds 300147 tons), followed by Cereals and Millets (195874) and Pulses (62931).

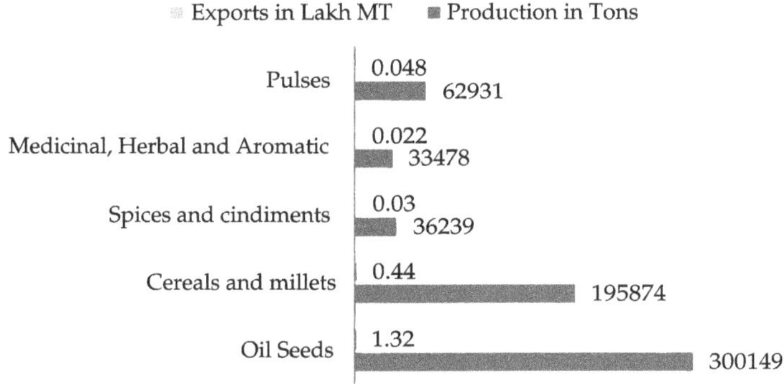

Fig. 2.5: Crops productions (Tons) and Export Quantity (Lakh MT)
Source: *APEDA Report, 2017.*

Nevertheless, most of the land area in India is used for grazing. Rest of the cropland used to cultivate organic agriculture like, cereals, green fodder, oilseeds, pulses & legumes, fruits & vegetables, tea & coffee, nuts, spices & condiments and herbal vegetation with medicinal values. Furthermore, excluding the agricultural cropping scope organic concept is further applicable in live stock farming in form of animal husbandry, fishery and bee hiving. Due to there is a close proximity exists between the soil and animals there should be minimum use of agro-chemical products in the agricultural land to balance the ecological harmony. Nowadays, to provide quality food to the health conscious consumer livestock farmers are also encouraged lowering the application of chemical drugs, antibiotics and artificial hormones for animal welfare and social wellbeing.

Government Initiatives to Boost Organic Farming in India

Government of India, also supported this movement through launching different schemes to boost organic farming in the country. Initiative like National Agro Forestry Policy-2014 , provides a platform to propel the agricultural farming through conservation of natural resources and protecting the environment. The policy, also has different objectives like supporting innovative technologies, publicity campaigns, quality assurance of bio-fertilizers, market research and development and also management of bio fertilizer, bio-control and waste decomposer organisms. Since 2001, Govt. of India has initiated a scheme like National Project on Organic Farming (NPOF) in 10th Five Year Plan and continuing as on date is one of the initiative to promote organic farming as well to provide certification established in different states of the country including, Ghaziabad, Bangalore, Bhubaneswar, Panchkula, Imphal, Jabalpur and Nagpur. They also support funding, through Capital Investment Subsidy Scheme (CISS) for production units for agro-waste compost, bio-fertilizers and bio-pesticides, establishing quality control regime and human resource development etc. A sum total of 56 nos. biofertilizer production units and 17 nos. of fruits and vegetables waste compost units have been established under NPOF.

Similarly, to promote organic farming and reducing dependence on chemical inputs in the process of cultivation, Government has launched a programme called as Rastriya Krishi Vikas Yojana (RKVY). Accordingly, to facilitate technical support and land expansion Agriculture Ministry also launched National Project on Promotion of Organic Farming (NPOF-DAC). To encourage farmers to participate in the certification process, they also introduced farmer group centric organic guarantee system under Participatory Guarantee System of India.

In addition to the above, National Horticulture Mission (NHM) and Horticulture Mission for North East & Himalayan States (HMNEH) also provides financial support to the organic farmers to set up vermicompost production units and to initiate organic farming; they approved Rs. 10, 000/- per hectare for maximum area of 4 hectare as well as certification cost at Rs. 5.00 lakh per 50 hectares to the farmers belong to the cluster. Furthermore, National Food Security Mission also supports awareness of Rhizobium culture and Phosphate Solubilising bacteria in the process of cultivation. To facilitate research and development in this area, a Network Project on Organic Farming (NPOF-ICAR) has been launched under Project Directorate of Farming System Research consisting of thirteen numbers of centres in India.

Sustainable Organic Farming: Scope and Challenges

In India, organic farming technique has been practiced since ages though application of natural resources as fertilizer and botanical pesticide as well. India is also tapping the global market for organic products significantly. Hence, organic sector has brought an excellent opportunity for its stakeholders of the supply including farmers, retailers, investors, marketers, agencies, importers as well as government. With the increasing percentage people interested for non-chemically affected products, the scope for organic farming in India is also booming. Different organic products accepted in global context are basmati rice, tea, coffee, oil seeds, cereals, spices, herbal medicines and also non-food products like cotton, jute and cosmetics, etc.

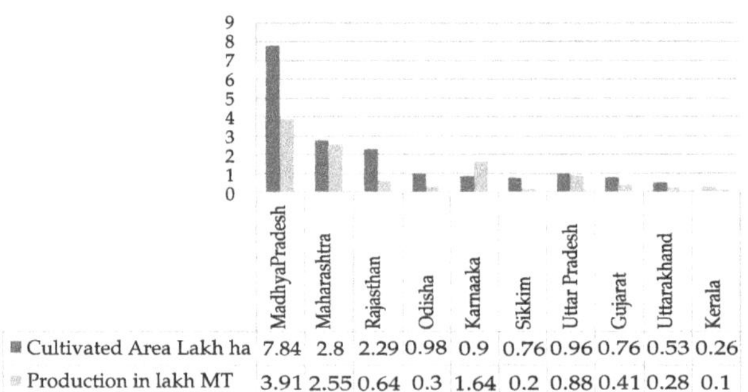

Fig. 2.6: Major States Contributing to Organic Farming in India in 2016-17

Source: *APEDA Report, 2017.*

According to Government of India report, 2015, organic food market would touch $1.36 billion by 2020. Likewise, as per TechSci Research report, 2015 the domestic market for organic food is expected to reach CAGR of 25% by 2020. In India, organic farming has been well adopted in top ten states like Madhya Pradesh, Maharashtra, Rajasthan, Odisha, Karnataka, Sikkim, Uttar Pradesh,

Gujarat, Uttarakhand and Kerala (Figure 6). In 2016-17 Madhya Pradesh has highest contribution towards the total organic farming scenario of the country. As depicted in Figure 6 , Madhya Pradesh has total cultivated area of 7.84 ha and total production of 3.91 lakh MT as on 2017.

Accordingly, Table 1 presents the growth status of state wise farming area in Hector from 2011 to 2017. As reported by Ministry of Agriculture and Farmer's welfare, there was 0.67 per cent raise in terms of organic certified land in Madhya Pradesh from 461774.726 hectare during 2015-16 to 464859.434 hectare in 2016-17.

Table 1.1 State wise Farming Area (in Hector) under Organic Certification from 2011-12 to 2016-17

Top Ten States /UTs	2011-12	2012-13	2013-14	2014-15	2015-16	2016-17
	Farming Area (ha.)					
Madhya Pradesh	382129.49	144239.76	60843.51	397546.62	461774.73	464859.43
Maharashtra	192434.33	66504.92	85536.66	134524.27	198352.29	224007.52
Rajasthan	63901.93	38289.05	66020.35	107523.24	155020.27	151609.91
Odisha	43787.24	18186.41	49813.51	81533.83	95896.98	92190.10
Karnataka	43815.43	27191.27	30716.21	52473.16	93963.34	81089.10
Sikkim	25408.55	43107.74	60843.51	76392.38	75851.21	75218.28
Uttar Pradesh	9798.76	32889.85	44670.1	53954.84	61081.83	56249.39
Gujarat	41978.94	45275.63	46863.89	49353.55	76813.06	64241.06
Uttarakhand	40547.95	20563.75	24739.46	36880.23	37221.39	30907.42
Kerala	15790.49	10568.41	15020.23	22980.9	25899.40	24812.78

Source: ***Retrieved from https://community. data.gov.in/state-ut-wise-farming-area-under-organic-certification-from-2011-12-to-2016-17.***

Despite its congenial climate and huge potentiality for organic cultivation, there is comparatively a lesser number of certified organic land in the greener state Kerala. However, in 2018 with 619 farm clusters, certified organic farm land reached 24812.78 hectares in all over Kerala.

On the other hand, among these above states Sikkim has excelled successfully through cultivating 75, 000 ha of certified organic agricultural land. This has led Sikkim declared as a fully organic state in 2016 through contribution of about 800, 000 tonnes of organic food products out of the 1.24 million tonnes of India's overall organic production. Accordingly, other states like Uttarakhand, and few tribal districts of Gujarat are also trying to follow the model adopted by Sikkim to achieve sustainable growth transforming them into 100% organic state. There has been many remarkable awareness programmes have conducted to sensitize farmers towards sustainable development through organic cultivation. NGOs and SHGs, also playing a major role to motivate farmers and consumers to adopt organic products. Besides, agro-eco tourism is another scope of organic farming where travellers can simultaneously enjoy the nature and get familiar with farming culture that further enhances the rural economy significantly.

Scope for Organic Agriculture in Odisha

Odisha is mostly an agrarian state having cropland area of 87.46 lakh hectares and where 73 % of populace depends upon agriculture for their livelihood. The topography of the state also facilitates the conducive environment for the development of organic cultivation in a tremendous way. Gradually, many farmers in Odisha have started adopting sustainable technologies to convert their farms in to organic from non-organic as well.

According to Ministry of Commerce and Industry report, during 2016-17 Odisha ranks 4 among top 5 States and UTs in terms of organic certified farming area of 92190.1 hectare which is clearly seen from the above Table -1.1 As presented in the below Graph (Figure-7), from 2011 to 2017 the growth rate of certified organic farm land has increased to 52.5 percent in Odisha state. On the contrary, organic farming scope in the state has slightly declined to -3.87 % in current year to 92190.1 from 95896.981 hectare in 2015-16. Although, there is rise in organic certified farming area to 14.97 % during 2016 in comparison to year 2014-15 .

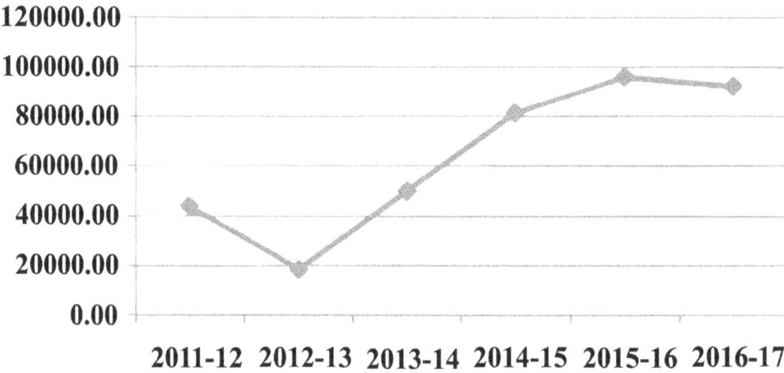

Fig. 2.7: Growth Status of Organic Cultivation in Odisha from 2011-2017

Source: *Retrieved from https://community.data.gov.in/state-ut-wise-farming-area-under-organic-certification-from-2011-12-to-2016-17/*

Correspondingly through involvement of around 40,000 farmers in organic harvesting the quantity of food production in organic sector rose to 33,994 tonnes per year. Mainly, the products certified for organic cultivation in the state are turmeric, cotton, cashew and vegetables. At present, the concept of organic farming has been included as one of the major objectives of the agricultural policy of Odisha. At present through Paramparagat Krishi Vikas Yojana (PKVY), Organic farming is being promoted in around 16,000 acres in 10 districts of the state. Government has embarked different initiatives to propel and support organic farming infrastructure of Odisha. United Nations Environment Programme (UNEP), recently joined hands with Odisha to promote organic cultivation via sustainable technology at tribal zones of the state.

Odisha State Seeds and Organic Products Certification Agency (OSSOPCA) have been developed by Agricultural Promotion and Investment Corporation of Orissa Ltd (APICOL) to drive organic cultivation in different districts of the state. Likewise, not for profit organization like Chetna Organic, Suvidha , Nirman,

Kisan swaraj and Sanjeevani are promoting and supporting sustainable farming in Odisha successfully. Similarly, India's popular trade promotion association like Associated Chambers of Commerce and Industry of India (ASSOCHAM) also established a national commission for organic farming on public – private participation mode. In collaboration with Agricultural and Processed Food Products Export Development Authority (APEDA), ASSOCHAM also proposed for initiating a training institute for the farmers in order to enhance their skill in the field of organic agribusiness. According to Business Standard Report (2018), a multi business conglomerate like ITC is planning to invest in organic cotton farming and aqua cultivation in Odisha .

Moreover, organic farming could generate huge employment opportunity from farming to allied services in a considerable manner. In 2017, Odisha played a major role to elevate country's share in global organic exports from about 0.2 per cent to about 2.5 per cent. Hence, Odisha definitely has an enormous potentiality to be India's export hub while developing a niche in the global market through export of certified organic herbs exclusively produced from the Odisha soil.

Issues and Challenges for the growth of Organic Farming in India

Undoubtedly, organic farming is the best method for sustainable agricultural prosperity of the country. However, the sector also faces many challenges for its successful implementation. Though India's agricultural traditions present the best suitable platform to flourish organic cultivation; non-organic farming has dominated this field since long days. One of the major challenges is to make both farmers and consumers aware of the benefits of organic cultivation over in-organic farming. Major of the strata highly rely on synthetic fertilizers and pesticides in their agro practices. Due to utilization of non-chemical inputs in the agricultural procedure, the productivity is comparatively slower; hence farmers aiming for a fastest production often rely on chemical based farming as it provides faster results.

Similarly, organically grown products have quite natural appearance and smaller in size. Hence, people easily attract towards chemically nurtured eye-catching agro products neglecting the high risk factors associated with them like pesticide toxicity and low contents of micro nutrients in the foodstuff. However, from few cases it was evident that organic cultivation are also prone to be contaminated from bacteria due to use of compost derived from animals. Similarly, due to the commercialized macrobiotic bio manures are costlier than chemical fertilizers; subsidy provision is highly required to convince traditional agripreneurs to switch to organic farming from conventional. Accordingly, pest control in biodynamic way is very intricate and time consuming too. India could be highly beneficial through exporting organic foods; but there is need for an essential requirement of efficient supply chain management system and competent marketing strategy.

Furthermore, due to organic products follow strict adherence to quality that ensure food safety are somewhat expensive than non-organic products available abundantly in the market. Hence, to convince customers to buy organic products like fruits, vegetables, pulses, spices, dairy products, etc. is quite challenging for the marketers in India.

Even though, awareness regarding organic philosophy exists among farmers, agripreneurs and policy makers; it requires more streamlined professionally managed system through efficient biotechnological research, development and training programmes.

Conclusion

Sustainable organic farming helps in maintaining the balance between earth and men in a holistic way through addressing different issues regarding, biodiversity loss, excessive nitrogen flow, climatic changes and health disorders. Therefore, the sole objective of organic farming is the overall wellbeing of plants, animal and people (Sriraj, 2015).

Furthermore, organic farming establishes a synergy between nature and human through bio-friendly techniques like recycling, crop rotation, balancing soil nitrogen, botanical pesticides and non-contamination of land and water resources. Presently, Indian market for organic products is growing at a robust rate with the increasing trend of eco-friendliness and health consciousness among people. As one of the oldest sustainable agricultural nation in this world, Indian organic farming sector has vast opportunity to accomplish global competitiveness and growth sustainability through strategizing innovation. However, apart from the healthiness point of view, it also has to prioritize on the development of rural livelihood and ecology.

Due to serving farming and saving farming is the latest mantra for the prosperity of the country, these above discussed challenges can be handled tactfully through effective utilization of manpower, technology, government initiatives and public-private partnerships .

Similarly, sustainable organic agriculture through a synergistic blending of traditional skill & knowledge along with cutting edge technologies results in making the evergreen India concept successful through empowering the rural socio-economic status along with ensuring food safety and security as well.

Lastly, adoption of an effective agribusiness model and marketing strategy for organic agriculture would definitely ensure wellbeing of farmers as well as healthy consumers (Kaur, 2012).

References

Akpinar, N., Talay, I., Ceylan, C., & Gündüz, S. (2005). Rural Women and Agrotourism in the context of Sustainable Rural Development: A Case Study from Turkey. *Environment, Development and Sustainability*, *6*(4), 473-486.

Bhattacharyya, P. & Chakraborty, G. (2005). Current status of organic farming in India and other countries. *Indian Journal of Fertilisers*, *1*(9), 111.

Chadha, S., Saini, J.P., & Paul, Y. S. (2012).Vedic Krishi: Sustainable livelihood option for small and marginal farmers. *Indian Journal of Traditional Knowledge* ,*11*(3),480-486

Das, B., & Pani, A. (2013). Agro-Tourism Development in India for Inclusive Growth and Global Competitiveness through Ideations and Innovations. *Indian Economic Journal, December,61-69*

Dubey, R. K., & Shukla, N. (2014). Organic farming: an eco-friendly technology and its importance and opportunities in the sustainable development. *International Journal of Innovative Research in Science, Engineering and Technology*, *3*, 10726-10734.

Eshun,G.and Tette,C.(2014). Agrotourism development in Ghana: A study of its prospects and challenges at Adjeikrom Cocoa Tour Facility. *Bulletin of Geography*. Socio–economic Series No.25, 81–99

Gahukar,R.T.(2009).Sustainable agriculture in India: Current situation and future needs. *International Journal of Agricultural Sciences*, *5*(1), 1-7.

Jaganathan,D ,Bahal Ram & Padaria R.N.(2008).Prospects of Organic farming in India: An appraisal . *Journal of Global Communication* , *1*(2) , 181-193

Jakovi , B., Tubi , D., & urovi , M. (2015). Sustainable development of rural tourism by developing new and authentic tourism products. *Tourism in Southern and Eastern Europe*, *3*, 103-114

Kesavan, P. C., & Swaminathan, M. S. (2007). Strategies and models for agricultural sustainability in developing Asian countries. *Philosophical Transactions of the Royal Society B: Biological Sciences*.

Kilcher, L. (2007). How organic agriculture contributes to sustainable development. *Journal of Agricultural Research in the Tropics and Subtropics, 89*, 31-49.

Mahindarathne, P. P., & Gunaratne, L. H. P. (2015). Entrepreneurial orientation of organic farmers: A case of organic vegetable farmers in the Badulla district of Sri Lanka. *Journal of International Food & Agribusiness Marketing*, *27*(4), 324-336.

Naresh, R. K., Shukla, A. K., Kumar, M., Kumar, A., Gupta, R. K., Tomar, V. S., & Singh, S. (2018). Cowpathy and Vedic Krishi to Empower Food and Nutritional Security and Improve Soil Health: A Review, *Journal of Pharmacognosy and Phytochemistry*, *7*(1), 560-575.

Niggli,U.(2015).Sustainability of organic food production: challenges and innovations. *Proceedings of the Nutrition Society.74* (1), 83-88.

Pandey,J. & Singh,A.(2012). Opportunities and constraints in organic farming: An Indian perspective. *Journal of Scientific Research* , *56* , 47-72

Ramesh, P., Singh, M., & Rao, A. S. (2005). Organic farming: Its relevance to the Indian context. *Current science*, *88*(4), 561-568

Reddy, B. S. (2010). Organic Farming: Status, Issues and Prospects-A Review. *Agricultural Economics Research Review*, *23*(2) , 343-358

Rigby, D., & Cáceres, D. (2001).Organic farming and the sustainability of agricultural systems. *Agricultural systems*, *68*(1), 21-40.

Roychowdhury, R., Abdel Gawwad, M. R., Banerjee, U., Bishnu, S., & Tah, J. (2013). Status, trends and prospects of organic farming in India: A Review. *Journal of Plant Biology Research*, *2*, 38-48.

Sharma, S. and Maheshwari, S. (2014).Constraints faced by tribal women in adoption of organic farming. *Advance Research Journal of Social Science, 5*(2):224-229.

Sowmya, K.S. (2014).Organic Farming- Scope in India. *Global Journal of Advanced Research, 1*(2) , 101-102

Vyavahare,P.J and Bendal,S. (2012).Transforming Agriculture in to Agriprenuership: A Case Study of Sitabai Mohite. *Asia Pacific Journal of Management & Entrepreneurship Research* , 1(2) , 164-173

Web Resources

Agriculture as an occupation losing lure in Odisha.(2015,September10).Retrieved from https://www.business-standard.com/article/news-ians/agriculture-as-an-occupation-losing-lure-in-odisha-115091000562_1.html

Behara,N.(2018,March 06). ITC keen to invest in organic farming, aqua cultivation in Odisha. Retrieved from https://www.business-standard.com/article/companies/itc-keen-to-invest-in-organic-farming-aqua-cultivation-in-odisha-118030600819_1.html

Govindarajan,V.(2014,February21). Transforming Rural India through Agricultural Innovation.

Retrieved from https://hbr.org/2014/02/transforming-rural-india-through-agricultural-innovation

Government of Odisha.(2015).Odisha economic survey 2014–15.Retrieved from http://www.indiaenvironmentportal.org.in/files/file/Odisha%20 Economic_Survey_2014-15.pdf

Growth Continues: Global Organic Market at 72 Billion US Dollars with 43 Million Hectares of Organic Agricultural Land Worldwide.(2015,February10). Retrieved from https://www.ifoam.bio/sites/default/files/ifoam_pr_2015_02_10_world_of_organic_agriculture.pdf

Lernoud and Willer (2018,March 22). The importance of organic agriculture for low/medium income countries-A statistical review. Retrieved from https://www.slideshare.net/francoisstepman/the-importance-of-organic-agriculture-for-lowmiddle-income-countries-a-statistical-overview

Mukherjee,S.(2018,March 28) . Organic food exports surge but certification remains a major issue. Retrieved from https://www.business-standard.com/article/economy-policy/organic-food-exports-surge-certification-remains-a-major-issue-118032800261_1.html

Noémi Nemes(2009).A Comparative Analysis of Organic and Non Organic Farming System. Retrieved from http://www.fao.org/3/a-ak355e.pdf

New govt scheme to promote organic farming in India. (2014,September 03). Retrieved from http://southasia.oneworld.net/news/new-govt-scheme-to-promote-organic-farming-in-india#.Vr3w8bR97cd.

Odisha Wakes up to Organic Farming.(2014,November 23).Retrieved from http://www.newindianexpress.com/states/odisha/2014/11/23/Odisha-Wakes-up-to-Organic-Farming/article2536958.ece

Odisha's organic farming can be worth Rs 23,000 cr in 5 years. (2012, September 13). Retrieved from https://www.thehindubusinessline.com/economy/agri-business/%E2%80%98Odishas-organic-farming-can-be-worth-Rs-23000-cr-in-5-years%E2%80%99/article20502028.ece

Organic Agriculture: At A Glance.(2015).Retrieved from http://www.krishijagran.com/farm/scenario-in-india/2015/03/Organic-Agriculture-At-a-Glance .

Organic farming can create 8 mn jobs in Odisha: Assocham.(2012, September 13). Retrieved from http://www.business-standard.com/article/economy-policy/organic-farming-can-create-8-mn-jobs-in-odisha-assocham-112091302007_1.html

Organic food market growing at 25-30%, awareness still low: Government.(2015 ,Oct.15). Retrieved from//economictimes.indiatimes.com/articleshow/49379802.cms?utm_source=contentofinterest&utm_medium=text&utm_campaign=cppst

Rao,R.(2015,August 28). Organic-Farming: A Boon or a Bane . Retrieved from https://www.business-standard.com/article/b2b-connect/organic-farming-a-boon-or-a-bane-115082800513_1.html

Sikkim becomes India's first organic state.(2016,January14).Retrieved from http://www.thehindu.com/news/national/sikkim-becomes-indias-first-organic-state/article8107170.ece

Sriraj,K(2015,June04). Challenges Organic Farming Face. Retrieved from https://www.dailypioneer.com/2015/columnists/challenges-organic-farming-faces.html

State/UT-wise Farming Area under Organic Certification from 2011-12 to 2016-17. (2018,August 14). Retrieved from https://community.data.gov.in/state-ut-wise-farming-area-under-organic-certification-from-2011-12-to-2016-17/

The Indian Organic Market: A New Paradigm in Agriculture .(2018,March 21), Retrieved from https://www.ey.com/Publication/vwLUAssets/ey-the-indian-organic-market-report-online-version-21-march-2018

Willer,H and Lernoud,J.(2015,February11).Organic Agriculture Worldwide: Current Statistics.Retrieved from http://orgprints.org/28216/7/willer-lernoud-2015-02-11-global-data.pdf

3

Sustainable Food and Livelihood Security: A Case of an Agricultural Intervention in Bangladesh

Ashish Jha

Introduction

Anderson (1990) defines the issue of food insecurity, as: "the limited or uncertain availability of nutritionally adequate and safe foods or limited or uncertain ability to acquire acceptable foods in socially acceptable ways". The issue of food security is more prevalent in rural communities. The challenges of food security and nutrition are worse for poor families. It is an undeniable fact that income is directly associated with food security and nutrition, if looked at micro level. It can be seen that family with decent income are out of the fear of food security and seldom experience any nutrition crisis. Various studies have established a direct link between food insecurity rates and household incomes.

Rural Communities, especially in low income countries are more vulnerable to food insecurity and lack of nutrition. Bangladesh is one of our neighbor countries, is also known for its issue of food insecurity. It stands at 83rd rank out of 113 countries in Food Security Index. doubt, the country is also facing poverty as a major challenge. During last decade, Bangladesh has shown remarkable improvements in terms of income and poverty.

To reduce poverty, once must have some livelihood opportunity. People need some livelihood activities to generate a basic income for their dignified survival. The researcher tried to look on the various livelihood programmes being taken by the government and non-government organisations in Bangladesh. During the study, the researcher came to know about a unique intervention carried out by a non-government organization in rural areas of Bangladesh.

The researcher carried out study of this programme in Bangladesh along with some case studies. The area of study is Barisal district under Barisal division of Bangladesh. The name of the programme is: *Sustainable Food and Livelihood Security* (SuFoL), and it is being implemented by Caritas Bangladesh.

Objectives and Methodology

The Objectives of this Paper are: (i) to understand the various sustainable agriculture practices being adopted in Bangladesh, (ii) to evaluate the impact of various on-farm and off-farm interventions on participants, (iii) to understand the role of Non-Governmental Organizations in promoting sustainable agricultural practices, and (iv) to disseminate the findings among the NGOs, institutions and individuals working for promoting sustainable agricultural practices in India.

This research paper is based on qualitative research design. Observation schedules were used to understand the various on-farm and off-farm activities. Interview schedules and case studies were used to assess the impact of the sustainable practices on the lives of the concerned people.

SuFoL

Sustainable Food and Livelihood Security (SuFoL) is a programme being run by a national non-government organization Caritas Bangladesh. The programme strives to provide sustainable livelihood to thousands of people in Bangladesh. This programme primarily revolves around agriculture and allied activities. SuFoL is being implemented in 72 villages of 14 sub-districts of Bangladesh. This programme was started in July 2013 and so far, assured and enhanced sustained livelihood status of around 3,664 beneficiaries. The researcher has done study in 5 villages under Barisal district.

Interventions under SuFoL: The interventions under this programme have been categorized under two heads - (i) On-Farm Intervention, and (ii) Off-Farm Intervention. Both the interventions target only marginal or landless farmers.

(A) On-Farm Intervention: It is a known fact that Bangladesh hosts frequent floods, thanks to more than 230 of the world's most unstable rivers crisscrossing it. Due to frequent floods, most of agricultural land in low lying plains remains flooded. This makes it almost impossible to cultivate those lands for almost 7-8 months in a year. Each year the situation is exasperated by flooding which restricts the time that crops can be grown. Floods affect over one million people in the country and more than 100,000 women, men and children are forced to move as villages and livelihoods are literally washed away. To cope up with this issue, the NGO offers financial and technical support for "Floating Garden". Floating garden is a concept through which vegetables can be cultivated over the flooded agricultural lands. This is very innovative intervention which makes agriculture, sustainable even during flood. A floating garden is built using aquatic weeds as a base on which vegetables can be grown. This extends the growing capabilities of rural communities, especially marginal farmers, who otherwise are unable to cultivate due to flood. This approach is amazingly cheap and sustainable.

How Floating Garden is Made? The first step to make a floating garden is to collect the water hyacinth. Floating water hyacinth is overlaid with bamboo poles of appropriate length to the size of the garden being constructed and this mass of plant matter is taken to one bank for further processes. Additional water hyacinth is collected and placed onto the bamboo layer and the thickness of the raft is built up. It is then woven into a buoyant raft. Once the raft's basic structure has been created the bamboo poles can be pulled out.

After around 7 to 10 days more water hyacinth is added to top up the existing structure then mulch can be added on to the water hyacinth base. Soil, compost and cow dung are added to cover the base of the raft to a depth of around 25 cm. The compost is made up from *azola*, good nitrogen fixing plant, and other readily available organic matter.

After this, seeds can then be sown. The technique used to improve the seeds' early development is to create round balls of compost comprising of decomposed water hyacinth and anorganic fertilizer known locally as *Tema*. A couple of seeds are planted into each ball and kept in a shaded area while germination takes place. Once the seedlings have begun to grow they can be planted out onto the raft.

***Crops can be Cultivated on Floating Garden*:** The floating garden is suitable for vegetable crops like Kang Kong (leafy vegetables), okra (lady's finger), gourd, brinjal, pumpkin, and onions.

How it is Cheap and Sustainable? As the floating garden is made by only naturally available materials, there is almost no cost, except labour. Further, after few uses, the raft will decay and can be broken up and used as compost and used in making new floating garden. It is 100% organic and does not leave any negative impact. Further, the decayed raft can be used as compost in land after flood, when rice is cultivated. This way it is sustainable.

(B) Off-Farm Interventions: SuFoL supports various off-farm income generation activities. The project targets female participants with no agricultural land as beneficiaries under these interventions. The major activities that are supported – financially and technically are Homestead based vegetable cultivation, poultry rearing, cattle rearing and vermi-compost preparing.The participants are given technical training on 'homestead based integrated vegetable cultivation and poultry rearing'. They are trained on raising and preparing garden beds, preparing compost and preserving seeds. They are also taught about duck, chicken, hen, goat and cow rearings. After successful completion of training, the participants are supported with initial amount to invest in any of these activities. The participants are provided with regular technical assistance till their business become sustainable. These interventions proved to be boon for many families, who were earlier struggling even for food to survive.

Case Studies

The researcher has taken two case studies of few beneficiaries of SuFoL programme. These are discussed below:

Case-I: Lagkhiram Minj – A Beneficiary of on-Farm Intervention

Mr. Lakhiram Minj (47) is a resident of Bakal village under Agailjaharaupazila in Barisal district in Bangladesh. He has studied upto 3rd standard. His wife Sreemoti Jarina (42) is a house wife. They have two sons, elder son Mr. Bibek Minj (25) is separately living with his wife and younger one, Mr. Nithanando Minj (21) is studying in class X. He has 2-acre land including homestead land one small pond. He used to produce single crop of paddy in a year from this land due to flooded water in his land. He was almost of no work during 7-8 months in a year. He was leading a miserable life, struggling to arrange food for his family. In June 2014, Caritas Bangladesh provided Lakhiram with training on vegetable production on a floating vegetable garden. He was given technical training on preparing floating garden, and provided with financial assistance of Tk1000. Since then, Lakhiram never looked back. He started doing vegetable farming on floating garden on his flooded land. His two sons use to help him. They all are now expert in Floating garden farming. Even people from neighbouring areas come to them and learn. He also crops rice once the flood die down. During the last flood, Lakhiram had enough food for his family. His family consumes fresh and organic vegetables, which eventually increased their nutrition status. During the last monsoon, Lakhiram was able to grow okra, kang-kong, and sweet pumpkin. There is little food in the markets during the monsoon, as few people can grow crops, so his vegetables are in great demand. He made a total of Tk 8000 during last monsoon. Lakhiram says "This has made a great difference to my life. Now I have enough food in the floods and I can give some to help my relatives as well and earn money during monsoon, when I was of no work earlier."

Case – II: Mrs. Morsheda – A Beneficiary of on-Farm Intervention

Mrs. Morsheda (47) is a widow and small farmer hailing from village of Bisarikati Charkhanda in the district of Barisal. She was dwelling in a hut, huddled in her mud house with her 4 daughters (all are married now). She didn't have own land for living and cultivation but only cash/rented land. Her husband (58) was a day labourer and died in March 2015. She started to work as a domestic worker in other's houses. She lived hand to mouth and could not manage meal for three times. Her earning situation was turning from bad to worse day by day. She always sought for any alternative sources of income but her effort ended in vain.

In December 2015, Morsheda was selected to take part in the Sustainable Food and Livelihood Security (SuFoL) Project in her village. She received training in 'Homestead based integrated vegetable cultivation and goat rearing'. She learned how to raise and prepare garden beds, prepare compost and preserve seeds. She also learnt the importance of using organic fertilizer and pesticides to improve crop production. Empowered with new skills, she got a support of Tk. 4500, with which she bought two Black Bengal Goats. After three years she is now the owner of eleven goats. She sold five goats Tk. 21,000 and she repaired her old house with Tk. 3,500. She also spent Tk. 3,500 for her grandson's education, cloths and Tk. 14, 000 for purchase one heifer. The present market price of the rest six goats about Tk. 25,500 and she has a plan to purchase a cow. Beside this she has cultivated vegetables such as cucumber, bottle gourd, red amaranth etc. in her homestead for her daily needs. Mrs. Morsheda is very happy with her initiatives. She is also using goat litters as manure in her kitchen garden. She quoted,"*I am very happy and grateful to the people for offering me support and cooperation for my success. I am confident that I would be able to contribute more for my family and society in future.*" Now the community people are coming to her for taking advice and suggestions on goat rearing. Obseving the success of Mrs. Morsheda, nine other farmers from nearby, have purchased black Bengal goat breeds and rearing on their backyard.

Conclusions

Sustainable Food and Livelihood Security (SuFol) is an exemplary micro level project ensuring food security and sustained livelihood opportunities to thousands of people. No doubt that such micro level on-farm and off-farm interventions are quite successful for atleast ensuring enough and nutritious food and some basic income. The floating garden is definitely that approach, which can be used to improve the food resources of people living and working not only in Bangladesh but also where flood or stagnant water is major problem. Such kind of agricultural interventions can make agriculture, a sustainable accord. SuFoL is successfully tackling various challenges like food insecurity, poverty, mal-nutrition, gender disparity and climate change adaption for agriculture. Such kind of projects should be studied and tried in India too.

References

Global Food Security Index (https://foodsecurityindex.eiu.com)

Caritas Bangladesh Website. (www.caritas.bd)

Food and Agriculture Organization (FAO)

4

Productivity and Efficiency of Asian Agriculture

Sonali Sharma and Kakali Majumdar

Introduction

Productivity is considered a method of measuring efficiency with which utilization of inputs in the production is carried out, other things being equal (Shafi, 1984). At an earlier time the term productivity and efficiency were used synonymously or can be used interchangeably (Afriat, 1972; Kaliranjan and Shand, 1985). However, they are not precisely the same things. Productivity in agriculture can be described as the ratio index in which we take total agricultural output to the index of total input which was employed in production (Pandit, 1965). Coelli (2005) defines the term "productivity" as total factor productivity (TFP) which is the magnitude that involves all the factors utilized in the production. Further the traditional productivity techniques, like fuel productivity in power station, productivity of labour in factory, and land productivity (yield) in farming etc. The productivity calculated considering single inputs are usually said to be the partial methods of productivity. Partial productivity methods can give a misguiding indication of productivity as a whole when ruminated in isolation (Coelli, 2005). Term productivity has been applied with different meaning and has provoked many conflicting interpretations. Sometimes it is considered as the overall efficiency with which a productive system works, while others it is taken as a ratio of output to resource expanded separately or collectively Efficiency of a production unit implies the assessment among the observed and the potential/ optimal output or input (Neogi and Ghosh, 1994).

Efficiency is the essential matter in every sphere of life and definitely true in economics. The efficiency in the process of agricultural production is commonly termed as agricultural productivity (Nishimizu & Page, 1982; Fare et al. 1989). Thus efficiency and productivity even if are not synonymous but highly interdependent. To understand clearly how these two concepts are interlinked the bifurcation of productivity is very important (Ludena, 2010). Productivity generally depends on two factors viz., type and quality of inputs and how these inputs are used in production process. The former characterizes the production technology while the second denotes the technical efficiency of the production process. Technical efficiency is referred to the producer's behaviour which concerns the production of output with a certain amount of inputs (Battese and Coelli, 1991). We can study the efficiency of production as a relationship among the observed and optimal value of its inputs and output (Cooper et al., 2004). Again efficiency is classified into three categories, namely: technical, allocative and scale (Battese, 1992). The term technical efficiency is already discussed. Allocative efficiency is related to the choice of the producer for the combination of optimal input (Kumbhakar, 1988). And finally, scale efficiency is a situation of choice of right quantity of output, where price and marginal cost of production are equal (Greene, 1993; Huang and Bagi, 1984).

The green revolution increased very quickly across Asia, and their subsequent growth in food production dragged the region back from the control of famine. Within 25 years agricultural harvests and outputs doubled over an era from 1965 to 1909 (Hazell, 2009). The WTO Agreement on Agriculture (AOA), which came into force in 1995, represents a significant step towards reforming agriculture and making it fairer and more competitive. The changes in agriculture trade policies in the post WTO context proved detrimental to most developing economies, especially in ASIAN region. One major reason is that a large number of these countries started re-Orienting their agriculture development policies towards growing non-food or commercial/high value crops (Viswanathan et al., 2012). In Asia, almost all the Asian countries are agriculture dominant and agriculture is the most important sector that offers occupation to a huge section of the population and the key to improve economic growth of the countries (Anthony, 2010). The contribution of agricultural sector in aggregate is on an average of 32 % to GDP throughout the previous 15 years. The highest, 32 %, is witnessed in Nepal and lowest, 1 % approx. in case of Japan.

Literature Review

In this section, the relevant literature has been reviewed to justify the objectives and methodology of the study. These reviews are summarized as follows:

A. Productivity and Efficiency Analysis of Asian Countries

Yujiro Hayami (1969) in his paper "Sources of Agricultural productivity gap among selected countries" tried to explore the cause of vast agricultural productivity differences prevailing between less developed and the developed countries. It aimed at identifying factors that define variances in the agricultural productivity of the less developed and developed countries. The main objective was to determine the factors which decide the farm production of countries by calculating the total production function of agriculture on the data of cross-country with the estimation of production elasticities differences in labour productivity among United States and India and Japan and India. In the result, three factors had been recognised that can impact the level of farm productivity. 1) Among different developed and less developed countries a main share of huge differences in the labour productivity of agriculture. 2) The difference in man-made modern factors accounts further productivity gap than variances in endowment of original factors. 3) Education and Research are important factors in shaping the productivity gap.

Toshihiko Kawagoe and Yujiro Hayami (1985) conducted an inter-country cross-section comparison of different countries of total productivity and agricultural production efficiency. The method used by Hayami and Kawagoe was a straightforward application of the total productivity index commonly used for the analysis of historical growth in national economics or section of economics. Variables used for the comparison were labour, land, livestock, fertilizers and machinery.

John Komlos (1988) criticized the work of Gregory Clark. Komlos argued that Clark was not able to compare the measurement of labour – productivity differences correctly Komlos also focused on the soil quality, fertilizers availability, horsepower employed, which were not similar in numerous countries, also the land to labour ratios were very different than what Clark said. Komlos figured out that output estimation by Clark was inaccurate.

Capalbo, Ball and Denny (1990) came up with the view of evaluating different methods which were been used for comparison of productivity in agriculture. According to Susan, Capalbo and Denny the productivity measurement is overall linked to the production technology. All the techniques of productivity measurement include the comparison of the level of output produced in relation to the inputs been taken.

Alexej Lissitsa and Martin (2005) analysed efficiency and TFP change of big agricultural projects during their shift to a market economy in Ukraine. The methodology used in the paper was Data Envelopment Analysis (DEA) and the Malmquist productivity changes index (MPI). Alejandro Nin Pratt, Bingxin Yu and Shenggen Fan (2008) compared agricultural total factor productivity growth in India and China. Supawat and Xiaobing wang (2012) in their paper aimed at conducting an inter-country study of agricultural productivity growth in transition countries of Europe and Asia. Their specific interest was on the direction and magnitude of productivity growth of their market reforms over different stages. Supawat and Wang adopted a non-parametric Malmquist index and used an output distance function for measurement of productivity growth.

B. Productivity and Efficiency Measurement

The composed works on the measurement of efficiency delivers two contending methods for estimating the relative efficiency across firms using best-practice frontier: i) non-parametric frontier approach; and ii) parametric frontier approach. The non-parametric frontier approach was the first method, developed by Debreu (1951), Farrell (1957) and afterwards explained by Banker, Charnes and Cooper (1984), Färe, Grosskopf and Lovell (1985) and others to measure the technical efficiency via estimating a production frontier.

Later on Winsten (1957) introduced a new econometric based approach named corrected ordinary least square (COLS). The approach was based upon the econometrics-based measurement of production frontier. However, the third approach was is based upon the parametric estimation of frontier and known as Stochastic Frontier Analysis (SFA) developed autonomously by Aigner, Lovell and Schmidt (1977) and Meeusen and Van Den Broeck (1977).

Stochastic Frontier Analysis (SFA) uses econometrics techniques to measure a stochastic non-deterministic production frontier. Stochastic data envelopment analysis (SDEA) had been introduced by Li (1998) and Cooper et al. (2004) via considering a chance constraint in the old-style DEA model. The SDEA models are used to split the random effects from the technical efficiency scores. Data envelopment analysis (DEA) is method for estimating the efficiency of firms called decision making units (DMUs) whose performance is regarded by numerous indicators (Chen and Zhu, 2003).

DEA institutionalizes the dimension of technical efficiency based on anticipated discrete piecewise frontier made up by a set of Pareto-efficient decision making units (DMUs). In all cases, these Pareto-efficient DMUs positioned on the efficient frontier, related to the others, lessen the usage of inputs given the outputs (input-oriented measure), or increase the outputs given the inputs (output-oriented measure) and described the best-practice or peer units within the sample of decision making units DMUs. These Pareto-efficient decision making units (DMUs) have a standard efficiency score of 1 that no single DMU's score can exceed. The DMUs which don't remain on efficient frontier are viewed as relatively inefficient and acquires a score somewhere in the range of 0 and 1. The efficiency score of each DMU can be clarified as the radial distance to efficient frontier.

To put it simply, DEA forms a non-parametric surface frontier over the data points to regulate the efficiency of each DMU comparative to this frontier. In the literature numerous mathematical programming models have been suggested (Cooper et al., 1994 and Cooper et al., 2004). Basically, each of these models pursues to form which of n DMUs decide efficient frontier. The specific DEA model employed prescribes the geometry of this surface.

Data and Methodology

The Concept of Total Factor Productivity (TFP): Total factor productivity TFP (t+1), is a measure of "average" product. TFP growth may be characterized as the change in TFP between period's t and t + 1;

$$\frac{TFP[t+1]}{TFP[t\}} = \frac{y^{t+1}/[x^{t+1}]}{y^{t}/[x^{t}]} \quad (1)$$

$$= \frac{y^{t+1}/y^{t}}{f[x^{t+1}/f[x^{t}]} \quad (2)$$

$$= \frac{A[t+1]}{A[t]} \quad (3)$$

In this form Total Factor Productivity consists the ratios of "aggregator functions," which is, an index number. In fact, if $y^t = A(t)x^t$ and $y^{t+1} = A(t + 1)$

x^{t+1} where both input and output are scalar, the index in (3) becomes TFP(t + 1)/ TFP(t) = $(y^{t+1}/y^t)(x^{t+1}/x^t)$, which is a very simple index number form of total factor productivity.

Malmquist Productivity Index: To estimate the country wise fluctuations in the Total factor Productivity growth of agricultural industry applying the non – parametric Malmquist Productivity Index (MPI) method. The important highlight of MPI approach is that it putrefies the TFP growth into two equally exclusive components i.e. technical efficiency change and technical progress. The TFP variation, measured by the MPI, among period's t and t + 1, can be described with the period t technology as;

$$M_0^t(x^{t+1}, y^{t+1}, x^t, y^t) = \frac{D_0^t(x^{t+1}, y^{t+1})}{D_0^t(x^t, y^t)}$$

In the same way, the MPI using period t+1 technology may be well-defined as;

$$M_O(x^{t+1}, y^{t+1}, x^t, y^t) = \frac{D_0^{t+1}(x^{t+1}, y^{t+1})}{D_0^t(x^t, y^t)} \sqrt{\frac{D_0^t(x^{t+1})}{D_0^{t+1}(x^{t+1})} \frac{D_0^t(x^t, y^t)}{D_0^{t+1}(x^t, y^t)}}$$

Fare et al (1994) says that the MPI formula can be inscribed as;

$$M_0^t(x^{t+1}, y^{t+1}, x^t, y^t) = \frac{D_0^t(x^{t+1}, y^{t+1})}{D_0^t(x^t, y^t)}$$

The first ratio on the right hand side of above equation evaluates the instabilities in technical efficiency amongst period t and t + 1 as a catching up to the frontier effect. The second term deals with the measurement of the change in production technology (i.e., technical change) generally signified as a change in production frontier.

$$M_O(x^{t+1}, y^{t+1}, x^t, y^t) = \frac{D_0^{t+1}(x^{t+1}, y^{t+1})}{D_0^t(x^t, y^t)} \sqrt{\frac{D_0^t(x^{t+1})}{D_0^{t+1}(x^{t+1})} \frac{D_0^t(x^t, y^t)}{D_0^{t+1}(x^t, y^t)}}$$

Data Design: To estimate the Malmquist indexes and TFP, time series data on 24 Asian countries from 2011 to 2016 was used. For the purpose of agriculture productivity and efficiency, the study included production of rice (one output), area harvested, potash fertilizer and nitrogen fertilizer (three inputs) which were obtained from FAO statistics website.

Inputs: Land comprises land cultivated with rice crops which were measured in hectares. The sources of information were the FAO stat website. Fertilizer consists of nitrogen and phosphate consumed in agriculture by Asian countries. The data on fertilizers was obtained also from the FAOSTAT website.

Results and discussions

The esteem more noteworthy than one speaks to a change of proficiency or profitability. Mean by and large specialized efficiencies (table 1), show a general positive pattern after some time for test nations. Notwithstanding, nations did not have same execution amid the period. Countries like China, Bangladesh, Cambodia, India, Tajikistan, Thailand, Turkey and Uzbekistan have encountered a major increment of in general specialized proficiency amid the period. Swinging to the Malmquist add up to factored profitability file table 2 incorporates mean estimations of proportions of progress in all out factor profitability list and its segments (proficiency change and specialized change). Means are specified for the example overall and also as by nation. Taking a gander at the example all in all, the adjustment in all out factor efficiency of horticultural segment of the nations has been certain.

Table 1: Mean Technical Efficiencies Change

Countries	Technical efficiency change (effch)	Pure technical efficiency change (pech)	Scale efficiency change (sech)
Afghanistan	0.994	1.001	0.993
Bangladesh	1.032	1.004	1.029
Bhutan	0.993	1.039	0.956
Cambodia	1.040	1.044	0.996
China	1.039	1.066	0.974
India	1.005	1.047	0.990
Indonesia	1.000	1.000	1.000
Iran	0.971	0.976	0.995
Iraq	0.979	0.986	0.992
Japan	0.963	0.970	0.992
Kazakhstan	0.979	0.989	0.990
Kyrgyzstan	0.974	0.984	0.989
Malaysia	0.922	1.028	0.897
Myanmar	0.897	1.000	0.897
Nepal	0.979	1.000	0.979
Pakistan	0.972	0.999	0.973
Philippines	0.957	0.975	0.981
Republic of Korea	0.822	0.837	0.982
Sri Lanka	1.000	1.000	1.000
Tajikistan	1.019	1.012	1.007
Thailand	1.002	1.000	1.002

Countries	Technical efficiency change (effch)	Pure technical efficiency change (pech)	Scale efficiency change (sech)
Turkey	1.014	1.012	1.002
Uzbekistan	1.009	1.008	1.001
Viet Nam	0.994	0.994	1.000

Source: *Own Estimation*

Table 2: Mean Total Factor Productivity Change

Countries	Technical efficiency change (effch)	Technological Change (techch)	Total factor productivity change (tfpch)
Afghanistan	0.994	0.833	0.828
Bangladesh	1.032	0.692	0.714
Bhutan	0.993	0.864	0.859
Cambodia	1.040	0.823	0.856
China, mainland	1.039	0.826	0.858
India	1.005	0.853	0.857
Indonesia	1.000	1.172	1.172
Iran (Islamic Republic of)	0.971	1.188	1.154
Iraq	0.979	1.167	1.143
Japan	0.963	1.186	1.142
Kazakhstan	0.979	1.213	1.187
Kyrgyzstan	0.974	1.162	1.131
Malaysia	0.922	0.877	0.809
Myanmar	0.897	0.937	0.840
Nepal	0.979	0.842	0.824
Pakistan	0.972	0.897	0.872
Philippines	0.957	0.89	0.851
Republic of Korea	0.822	0.837	0.687
Sri Lanka	1.000	0.809	0.809
Tajikistan	1.019	0.864	0.881
Thailand	1.002	0.848	0.850
Turkey	1.014	0.851	0.863
Uzbekistan	1.009	0.851	0.859
Viet Nam	0.994	0.759	0.754

Source: *Own Estimation.*

Conclusion

In this article, the relative execution of the horticultural segment was measured utilizing DEA from a board informational index of 24 nations, including a multiyear time span from 2011-2016. Scientific programming strategies were utilized to gauge Malmquist files of aggregate factor profitability encountered a positive development in the Asian nations like Indonesia, Iran, Iraq, Japan, Kazakhstan, Kirgizstan add up to factor efficiency scores were recorded, which demonstrates the most noteworthy agrarian efficiency than the other Asian nations considered in this examination. The main thrusts were changes in EffCH and TechCH. Whatever is left of nations considered in this examination were the minimum proficient nations in the example. These outcomes have significant impacts for policy directing. The fundamental obstacle over the long run lies in the slow rise in technical change. This specifies that there is an increasing firmness for continuous enhancements of technology, which need energetic character for the public sector in alliance with agriculturalists or farmers to increase the technology level expressively over time.

References

Afriat, S. N. (1972). Efficiency estimation of production functions. *International Economic Review*, 13(3), 568-598.

Aigner, D., Lovell, C. K., & Schmidt, P. (1977). Formulation & estimation of stochastic frontier production function models. *Journal of Econometrics*, *6*(1), 21-37.

Anthony, E. (2010). Agricultural credit & economic growth in Nigeria: An empirical analysis. *Business & Economics Journal*, *14*, 1-7.

Banker, R. D., Charnes, A., & Cooper, W. W. (1984). Some models for estimating technical & scale inefficiencies in data envelopment analysis. *Management science*, *30*(9), 1078-1092.

Battese, G. E. (1992). Frontier production functions & technical efficiency: a survey of empirical applications in agricultural economics. *Agricultural economics*, *7*(3-4), 185-208.

Battese, G. E., & Coelli, T. J. (1992). Frontier production functions, technical efficiency & panel data: with application to paddy farmers in India. *Journal of productivity analysis*, *3*(1-2), 153-169.

Capalbo, S. M., & Ball, V. E. (1990). International comparisons of agricultural productivity: development and usefulness. *American Journal of Agricultural Economics*, *72*(5), 1292-1297.

Charnes, A., Cooper, W. W., & Rhodes, E. (1978). Measuring the efficiency of decision making units. *European Journal of Operational Research*, 2(6), 429-444.

Charnes, A., Cooper, W. W., Lewin, A. Y., & Seiford, L. M. (Eds.). (2013). *Data envelopment analysis: Theory, methodology, and applications*. Springer Science & Business Media.

Chen, Y., & Zhu, J. (2003). DEA models for identifying critical performance measures. *Annals of Operations Research, 124*(1-4), 225-244.

Coelli, T. J., Rao, D. S. P., O'Donnell, C. J., & Battese, G. E. (2005). *An introduction to efficiency and productivity analysis*. Springer Science & Business Media.

Cooper, W. W., Seiford, L. M., & Zhu, J. (2004). Data envelopment analysis. In *Handbook on data envelopment analysis* (pp. 1-39). Springer, Boston, MA.

Debreu, G. (1951). The coefficient of resource utilization. *Econometrica*, 19, 273–92.

Fan, S., & Chan-Kang, C. (2005). Is small beautiful? Farm size, productivity, and poverty in Asian agriculture. *Agricultural Economics, 32*, 135-146.

Färe, R., Grosskopf, S., Logan, J., & Lovell, C. K. (1985). Measuring efficiency in production: with an application to electric utilities. In *Managerial issues in productivity analysis* (pp. 185-214). Springer, Dordrecht.

Färe, R., Grosskopf, S., Lovell, C. K., & Pasurka, C. (1989). Multilateral productivity comparisons when some outputs are undesirable: a nonparametric approach. *The review of economics and statistics*, 90-98.

Farrell, M. J. (1957). The Measurement of Productive Efficiency. *Journal of the Royal Statistical Society. Series A (General)*, 120, 253-290.

Ganguly, N. (1987). *Rural Upliftment through Land Reform Measures*, Kurukshetra, November, 19-20.

Greene, W.H. (1993). The Econometric Approach to Efficiency Analaysis, in Fried, H.O.,Lovell, C.A.K. & Schmidt, S.S.(Eds), *The Measurement of Productive Efficiency*, Oxford University Press, New York, 68-119

Hayami, Y. (1969). Sources of agricultural productivity gap among selected countries. *American Journal of Agricultural Economics, 51*(3), 564-575.

Hazell, P. B. (2009). Transforming agriculture: The green revolution in Asia. *Millions fed: Proven successes in agricultural development*, 25-32.

Huang, C. J., & Bagi, F. S. (1984). Technical efficiency on individual farms in northwest India. *Southern Economic Journal*, 51(1), 108-115.

Johnson, M., Hazell, P., & Gulati, A. (2003). The role of intermediate factor markets in Asia's green revolution: lessons for Africa? *American Journal of Agricultural Economics, 85*(5), 1211-1216.

5

Climate Change and Sustainable Food Security

Navneet Kaur Nijjar

Introduction

Climate change is an issue hovering as a major roadblock to global development. It is a topic that worries the developed, the developing and the underdeveloped countries equally. While it is said that climate change is the result of the development process initiated by the developed countries during the industrial revolution, its outcomes shall be faced by every part of the world. Climate change has been a hot topic in recent international negotiations as well as is an important part of every country's development agenda. Erratic weather changes, changing patterns of precipitation, changing habitats, consumption patterns, adaptations, etc can be seen as direct or indirect effects of climate change. While this climate change affects directly on the climate and weather pattern of a place, its indirect effects can be seen over different sectors as spill over effects. NASA has reported that there has been a significant increase of about 0.9 degree Celsius since the second half of 19th century. Also, it notes that the major contributors to this increasing temperature are the higher concentration of greenhouse gas emissions and other anthropogenic activities.

The previous 40 years have been the warmest with five warmest years since 2010. Out of these, 2016 was the warmest year in which eight out of twelve months were the warmest on record. The melting glaciers are a concern as well which can be seen from the data from NASA's Gravity Recovery and Climate Experiment which says that Greenland lost an average of 281 billion tons of ice per year between 1993 and 2016, while Antarctica lost about 119 billion tons during the same time period. In addition to these there is also considerable evidence of climate change reflecting in rising sea levels, ocean acidification, and increase in occurrence of extreme events, glacial retreat, etc. Talking in particular about India, climate change has been attributed to the development process using conventional methods, increasing demand and increased energy generation, higher vehicular exhaust causing declining air quality, deforestation and changing land use pattern (FAO, 2008).

Climate change is also likely to pose a threat to food security. It is estimated that the changing climatic variability and weather will cause distortions in not only terrestrial ecosystem but will also disturb the marine ecosystem and biodiversity. Especially the ecologically sensitive zones will face greater challenges. The extinction of species due to their habitat loss is increasing. The marine ecosystem is also facing redistribution of fisheries. For many crops the threat will need to focus on adaptation measures. For example for areas producing crops like maize, wheat or rice, climate change of about 2 degrees Celsius or above will negatively affect the yield. "Climate change is projected to reduce renewable surface water and ground water resources in most dry subtropical regions. Until mid-century, projected climate change will impact human health mainly by exacerbating health problems that already exist", IPCC notes. Throughout the twenty first century, global climate change is anticipated to guide to will increase in ill-health in several regions and particularly in developing countries with low income, as compared to a baseline while not global climate change. By 2100 the mix of warm temperature and wetness in some areas for elements of the year is anticipated to compromise common human activities, as well as growing food and dealing outdoors.

In urban areas global climate change is projected to extend risks for individuals, assets, economies and ecosystems, as well as risks from heat stress, storms and extreme precipitation, landlocked and coastal flooding, landslides, pollution, drought, water inadequacy, water level rise and storm surges. These risks square measure amplified for those lacking essential infrastructure and services or living in exposed areas. Rural square measures are expected to expertise major impacts on water accessibility and provide food security, infrastructure and agricultural incomes, as well as shifts within the production areas of food and non-food crops around the world. Aggregate economic loses accelerate with increasing temperature, however international economic impacts from global climate change impacts are difficult to estimate.

From a poverty perspective, global climate change impacts are projected to prevent economic process, build poverty reduction harder, additional erode food security and prolong existing and make new poverty traps, the latter notably in urban areas and rising hotspots of hunger. International dimensions like trade and relations among states are necessary for understanding the risks of global climate change at regional scales. Global climate change is projected to extend displacement of individuals. Populations that lack the resources for planned migration expertise higher exposure to extreme weather events, notably in developing countries with low income. Global climate change will indirectly increase risks of violent conflicts by amplifying well-documented drivers of those conflicts like poverty and economic shocks.

Droughts and cyclonic storms have become frequent in many parts of our country. The seasonal rainfall in monsoon shows a decline of around 6 % to 8 % in eastern Madhya Pradesh, north eastern India and parts of Gujarat and Kerala over past century (Lal et al., 2010). On the other hand, storms have become a regular phenomenon in parts of Gujarat and West Bengal (Sehgal, 2009). Rising sea levels, deep rainfall, enormous soil erosion and landslides are posing another threat. It has been predicted that the adverse impact of climate change will be seen on developing countries as compared to the moderate effect on the developed countries. This has a serious concern for a country like India which has just been setting conditions for the take off stage to development (Rostow's stages of growth). The impacts are already been seen in terms of higher greenhouse gas emissions, pollution, health issues, increased occurrence of floods, droughts, etc. which disrupt the normal functioning of the people in addition to the additional costs it puts on the public treasury to combat the negative spillover effects of the so called "development activities". Further, the effects will be seen on every aspect of human life directly or indirectly.

Literature Review

India is blessed with its diverse geographical location and climatic topography. Given this fact, India produces a variety of crops across different regions. Climate change has affected different regions differently causing adverse effects on the productivity as well as nutrition of the produce. This has raised a question on the food and nutritional security for our growing population. India is on the path of economic growth and development with its positive and young demographic profile and we cannot afford to lose such an opportunity. Hence, it is important that any roadblocks in terms of climate change are checked and are resolved so as to pave the way for development. It has been seen that different regions are facing different problems caused due to climatic changes. For example, the western part will be affected by frequent drought situation while the eastern part will be facing regular floods leading to loss of productivity. Also, the northern India will be facing drop in their water-table threatening the crop productivity.

A study carried out by S. Naresh Kumar, P. K. Agarwal and their co-authors estimates that wheat production will decline by about 6 % to 23 % by 2050 and further by 15 % to 25 % by 2080 due to disturbances in sowing pattern as a result of climate change. Also, they said that the central and southern regions will face much adverse effects. In addition to this, Indian agricultural scientists have also found effects of climate change on rice productivity. 'Down to Earth' magazine reported an article featuring the study conducted by researchers at Coimbatore based Tamil Nadu Agricultural University on the rice production which is a staple food of the region providing 23 % of global human per capita energy and 16 % of global human per capita protein (International Rice Research Institute). It was observed that high temperature caused negative impact on the productivity of rice even if positive effects of high level of carbon dioxide were considered.

In a study by Harish Thakare and Prakash Kumar Srivastav, an attempt was made to investigate the cause of poor production of cotton in 2011-12 in Surat, Gujarat as compared to 2012-13 and 2013-14. The study adopted analysis of weather data from 2000 and compared it with the weather data of 2011-12. The results found that during 2011 there was a significant decline in the cotton productivity as there was delayed monsoon during the sowing stage and high rainfall during the flowering stage. This extreme weather patterns affected the cotton productivity. This process hampered the growth of the cotton crop and resulted in low yield as well as quality. The productivity of rain-fed pulses is commonly vulnerable by temperature changes thanks to the outstanding heat and drought particularly throughout pudding stage. Majority of the pulses growing regions are prone to temperature change as most threshold temperature for tolerance of pulses has already been reached on the far side at 35 degrees Celsius. Pulses are unable to derive edges of high carbon dioxide fluxes, as a result of these crops are full grown underneath climatologically stressed setting. Since each drought and heat are invariably perennial development in pulse growing regions, factor exploration from needed germplasm for stress tolerance would stay the possible future hope for evolving newer varieties tolerant to multiple abiotic stresses. Therefore, methods are created to develop climate-resilient varieties primarily based upon screening of enormous variety of germplasm. Early flowering, short length, quicker biomass accumulation, deep root-system, high water-use efficiency and high root proliferation before onset of terminal drought and warmth are found to be the specified methods to flee the abiotic stresses. variety of genotypes are known that are tolerant to heat stress supported high fertility of spore and pod formation at temperature on the far side of 40 degrees Celsius.

Under dynamic climatical condition, the soil organic carbon is subject to depletion and adoption of resource conservation technologies and management practices have tremendous potential in sequestering carbon in soils. Resource-conservation technologies, conservation-agriculture practices and improved farming practices will scale back emission of greenhouse gases (GHGs) and enhance carbon storage in soils. Moreover, organic system of pulse production will increase soil organic matter and cultivation of canopy crops and lessens the emissions of gases through production and transportation of artificial fertilizers thanks to less demand for them. The inclusion of pulses in crop rotation reduces the necessity for fertiliser inputs and contributes nitrogen to succeeding crops. Lesser fertiliser needs are considerably lowering GHGs emissions.

The effect of soil erosion on the productivity is also a significant concern. The excess soil erosion has been affecting the fertility of the soil as well as the salinity of the oceans. The changes in the hydrological cycle, rainfall patterns and their duration as well as magnitude will affect the rate of soil erosion causing question on food security. Indian institute of Remote Sensing in 2015 reported that about 147 million hectares of soil erosion. Soil erosion caused by water is at 56% while wind erosion accounts for 26% mainly in Asia and Africa (Oldeman, 1991). "In India, it is estimated that water erosion causes damages on 113 million hectares which threatens productivity and the fertility of the soil" (Kumar, 2004). It leads to reduction in the quality of land, loss of topsoil and decrease in the content of soil organic matter and thereby the loss in crop yield as it relates to high run off rates causing reduction in the water content and moisture holding capacity of the soil (Lal, 2001).

Climatic changes and its effects will soon take over as the biggest challenge if we do not act now. Many countries have recognised the concern for climate change and are moving towards mitigation and climate-smart technologies in different fields so as to combat the ill-effects of changing climate. The question of food insecurity is worrisome. And especially for a developing country like India it's a major cause of concern. To choose between development and environment is a tough choice which definitely involves many stakes but it's high time to act. As we see the whole country is seeing disturbed trends of crops and agricultural production because of erratic weather and distorted atmospheric conditions which also include the changing characteristics of soil, water and air. The National Action Plan for Climate Change analyses changing scenario and recommends to adopt sustainable management practices in terms of agriculture, energy generation, urban planning, managing forests, communities, etc. We need to dig deeper into these mechanisms so as to get positive and resilient results.

While there is a growing debate on the question of ethical issues on the development process and whether we are digging graveyards for our own selves, I would like to point the effect of climate change on some important aspects of food security issues and how we can adopt strategies to mitigate the effects of climate change. The paper presents some views about the mechanism of soil and ground water to help combat the ill-effects of changing weather patterns and increasing CO_2 emissions. Sustainable agricultural practices are gaining importance due to their resilient and efficient benefits. I would point to two important factors contributing to sustainable agricultural practice: soil and groundwater.

Discussion

The soil is the solution to the climate change. Soil is home to countless species that naturally recycle waste to value. Soil purifies water and is a natural reservoir of atmospheric carbon. Soil is often regarded as a living universe under our own feet. The carbon levels in atmosphere are at 492 parts per million as of 2015. Climate scientists argue that atmospheric levels of carbon dioxide must be decreased to 350 parts per million for our continued survival. The stress is laid on land management, agricultural and farming practices. The question is how we farm? Soil characteristics are affected by our tillage, amount of pesticides, management of waste, crops, etc. In many regions, excess use of fertilisers to earn higher value for more productivity has degraded the original composition of the soil. The nutrient value of the soil gets lost as the fertilisers have harmful chemicals that destroy many organic species that are essential for the maintenance of a healthy soil.

As the organic matter in the soil decreases the porosity as well as the water holding capacity of the soil is negatively affected. This further affects the quality as well as the quantity of the produce. The organic matter as well as the living organisms in the soil absorb atmospheric carbon and utilise it to nurture the growth of the plants. But when the organic matter decreases, the capacity of the soil to absorb atmospheric carbon dioxide also decreases. Carbon is not our enemy. It is the building block of life. But when it is not put to productive use and remains in the atmosphere it harms our ecosystem in form of greenhouse gases which acts as a good heat absorbing agent and thereby causes global warming. The question is that how to absorb excess carbon dioxide from the atmosphere. The answer is *Carbon Farming* which includes applying a thin layer of compost on the top layer of soil. This practice can absorb excess carbon and feed the microorganisms which in turn improve the texture and nutrient formulation of the soil. The target should be the regeneration of soil to capture climate change. This regenerative agricultural practice will not only make our soils healthy but also make our food healthy and nutritious. It is very important to go back to our traditional practices to arrest changing weather patterns. In addition to this, managing waste and converting it to value should be adopted, tilling should be monitored as well as grazing should be planned. The strategy should include multidisciplinary approach in forms of engineering, bio-technology and agri-science which should also be efficient and cost-effective. The investment in our soils is long term but is sustainable and hence effective.

Groundwater is another factor which is bearing the brunt of climate change. For an agrarian country like India, whose majority of agriculture is rain-fed, the situation is grave. The Green Revolution brought a wave of happiness among Indian farmers in terms of higher productivity and higher profits. But little did they know that their short term profits will become their long term regrets. The genetically modifies seeds ensured higher productivity at the cost of pesticides harming the soil as well as their higher utilisation of water creating downfall of underground water table and hence resulting in water stress. Ground water is used as store house of water which can be used in years of less rainfall and droughts. But the over-utilisation of water has led to exploitation of water reserve.

The situation is much grave in parts of northern and western India. The lack of nutrients and organic matter in the soil fails to absorb the rain water which flows causing soil erosion. This double edged sword does not recharge ground water pus it erodes the top layer of soil. With about 4 per cent of the water resources of the world, India should have been a water-adequate nation. But the experts say that in 2011, India became a water-stressed country. 210 billion cubic metres is the amount of ground water extracted in India annually, which is the highest in the world, Down to Earth reported. Ground water today provides for more than 60 % of net irrigated area. As a result, over 60 % of districts in the country are facing problems related to poor quality and shortage of ground water. It is expected that the demand for water will further rise drastically to about 833 billion cubic metres (BCM) in 2025 from present 712 billion cubic metres and further to 899 BCM in 2050.

World Resources Institute use online tool to evaluate water risk. The maps below show the grave condition of water resources in Indian states.

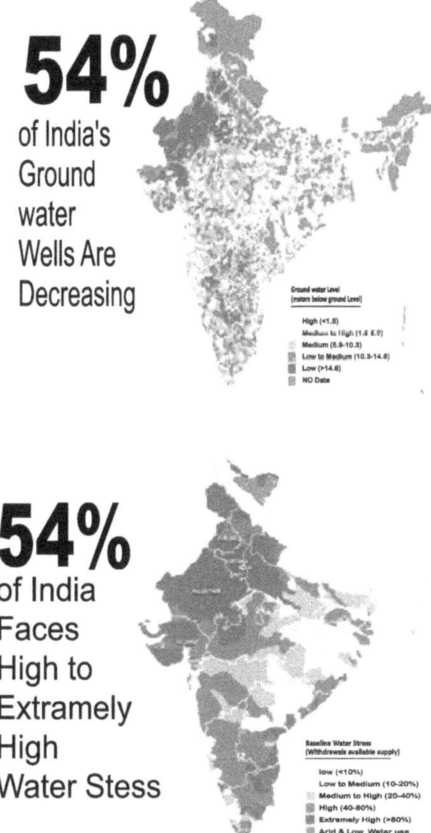

The Composite Water Management Index released by NITI Aayog says that around 600 million Indians face high-to-extreme water stress with about 2,00,000 people dying every year due to poor access to safe water. The situation is further expected to deteriorate as the population and demand for water will rise by 2050. NITI Aayog has called for urgent and improved management of water resources. The growing scarcity can also affect food security. If mitigation measures are not implemented, the GDP losses are estimated at approximate 6 per cent by 2050, the NITI Aayog report said. Water is an essential component of human life. The Index highlights that nearly 70 percent of water is contaminates which ranks India at 120 from 122 countries on global water quality index. These figures are alarming as Indian agriculture is highly dependent on the quality as well as quantity of water. This can challenge food security of states that are dominated by agriculture such as Uttar Pradesh, Haryana rank fairly low in the water quality Index. Further these states account for about 20 to 30 percent of India's agriculture. India's potential to increase its net cultivated area is huge using micro-irrigation of about 140 million hectares. This has not yet been reached with drip irrigation covering 3.37 million hectares, sprinkler irrigation covering 4.36 million hectares and totalling to 7.73 million hectares.

Conclusions

The mitigation measures to recharge ground water are of urgent need. Regeneration of groundwater should undertake region specific approach. Watershed development, water storage are useful measures for hilly areas. For example, in Dahod district of Gujarat, which is a dry and hilly region, watershed development programmes has been adapted for efficient utilisation of water. Agricultural practices should be analysed based on sowing crops according to the water table and the magnitude of rainfall of the region so that groundwater is not exploited. The water-use efficiency of the region should be taken into consideration. A report released by NABARD cites 10 major crops across the country and recommends realignment of cropping patterns to water availability (Water Productivity Mapping of Major Indian Crops). Rice, wheat, maize, red gram, chick pea, sugarcane, cotton, groundnut, rapeseed, mustard, potato – these occupy more than 60% of country's gross cropped area. Punjab reports highest land productivity for rice at 4 tonnes per hectare. But it produces only 0.22 kg of rice for every meter cube of irrigation water. On the other hand, Jharkhand produces 0.45 kg and Chhattisgarh produces 0.68 kg for every meter cube of irrigation water. The low productivity in these states is due to low irrigation coverage (Jharkhand has only 3% land under irrigation). Sugarcane, a water guzzling crop, Tamil Nadu reports highest land productivity at 105 tonnes per hectare followed by Andhra Pradesh. All these states are water stressed having irrigation productivity of less than 5 kg per meter cube. Tamil Nadu needs an average of 40 rounds of irrigation while the gangetic plains of Uttar Pradesh and Bihar need 5 and 8 rounds of irrigation respectively. Looking at these patterns, it is important that the cropping patterns be aligned to water availability using both demand and supply interventions. With water and power subsidies skewing crop patterns, it recommends reforms with shift from price policy approach of heavily subsidizing inputs to an income policy approach of directly giving money to framers on basis of per hectare. The price will then be determined by market forces.

Climate change mitigation measures will require long term efforts and strategies to see its results. But it is very important to act now as not taking measures will cause detrimental effects on every sector of our economy as well as pose a threat to our ecosystem. Food security is very crucial aspect for a developing country like India and for that we have to make sure that our agriculture is sustainable and the practices we adopt do not harm the environment further. For example, many regions have adopted rice and fish farming which include simultaneous production and nurture of rice and fisheries. it is important to find ways that incorporate our traditional farming methods and give them modern supplements of technology and knowledge. The importance of knowledge of the geography, climate, soil quality, ground water level etc. should be properly analysed and then it should be taken up for production of suitable crops. The components of geography, biotechnology, agriculture and economics should be applied together to make agriculture climate smart and resilient to future needs. Proper dissemination of knowledge about farming, crops and all other variables should be taken up. The basics should start with the land management measures and restoring groundwater levels for a sustainable future. This will also help farmers to increase their income and contribute to sustainable farming. Ensuring healthy soils and groundwater level will not only ensure the quantity of the crops but will also ensure quality and nutrient value of the crops. India, being an agrarian economy can adopt these measures and lead by an example.

References

Akoijam, S.L.S. (2011). Food Security: Challenges and Issues in India. *The International's Research Journal of Economics and Business Studies* 1(1), 1-11

Basu, P.S., Singh, U., Kumar, A., Praharaj, C.S., Shivran, R.K. (2016). Climate Change and Its Mitigation Strategies in Pulses production. *Indian Journal of Agronomy* 61(4), 71-82.

Brahmanand, P.S., Kumar, A., Ghosh, S., Chaudhary, S. Roy., Singandhupe, R.B., Singh, R., Nanda, P., Chakraborty, H., Srivastava, S.K., Behera, M.S. (2013). Challenges to Food Security in India. *Current Science* 104 (7), 841-846.

Brevik, E.C. (2013). The Potential Impact of Climate Change on Soil Properties and Processes and Corresponding Influence on Food Security. *Agriculture* 3(3), 398-417.

Dev, S.M., Sharma, A.N. (2010). Food Security in India: Performance, Challenges and Policies. *Oxfam Working Paper Series*, Number 7, 2-15.

Gupta, N., Mishra, A., Agrawal, N.K., Sathapathy, S. (2018). Interstate Cooperation for Climate Change Adaptation in Indian Himalayan Region. *Economic and Political Weekly* 53(12), 35-40.

Himanshu, (2013). Poverty and Food Security in India. *ADB Working Paper* Series 369, 2-12.

Kattumuri, R. (2011). Food Security and the Targeted Public Distribution System in India. *Asia Centre,* Working Paper 38, *London School of Economics and Political Science*, 5-23.

Kaur, P. (2014). Food Security in India- Some Issues and Challenges. *Excel International Journal of Multidisciplinary Management Studies* 4(11), 12-22.

Khan, H., Hasan, A. (2017). Climate Change: Concern for Food Security in India. *IOSR Journal of Humanities and Social Science* 22(10), 52-57.

Kumar, S.N., Agarwal, P.K., Swarooparani, D.N., Saxena, R., Chauhan, N., Jain, S. (2014). Vulnerability of Wheat Production to Climate Change in India. *Climate Research* 59 (3), 173-187.

Nanda, M. (2018). Addressing India's Food Security: Is Phosphorous the Missing Link? *Economic and Political Weekly* 53(15), 17-19.

NRAA (2011). Challenges of Food Security and Its Management 2011. *National Rain Fed Authority of India*. Planning Commission of India, New Delhi, Position Paper No.5.

Thakare, H.S., Shrivastava, P.K., & Bardhan, K. (2014). Impact of Weather Parameters on Cotton Productivity in Surat (Gujarat), India. *Journal of Applied and Natural Sciences* 6(2), 599-604.

Tripathi, A., Mishra, A. (2017). Farmers need more help to Adapt to Climate Change. *Economic and Political Weekly* 52(24), 53-59.

Upadhyay, R.P., Palanivel, C. (2011). Challenges in Achieving Food Security in India. *Iranian Journal of Public Health* 40(4), 31-36.

6

Efficiency of Agricultural Practices in Una District, Himachal Pradesh

C. Prakasam and R. Saravanan

Introduction

Agriculture can be influenced by numerous factors, drought put an extremely enter a job in it. Una is located in the outskirt of Himachal Pradesh state in the south-west side, whose essential occupation is farming with a topographical area of 1550 km^2, with a woods front of 523 km^2 different densities. The elevation of the region ranges from 300-600 meters for very nearly 75 % of the district and the remaining among 600-900 meters from the ocean level. A couple of edge tops and pinnacles additionally have elevation in excess of 900 meters. River Swan, which is the essential wellspring of water, is bolstered just by precipitation source in Shivalik lower regions spanning up to 55 kilometers. The river overflows and disintegrates the banks during a rainstorm and makes harms property and verdure lying close to the outskirts of the river. Being an agro land, Una is inclined to drought every now and again and the impacts of drought including the loss of flora, fauna and human lives, unhealthiness, crop disappointment, and water shortage. There were twenty-two major drought events during the 1901-2010, ranging from Moderate to Severe. The main drought-prone areas in Una district are Haroli and Bangana Tehsil. To mitigate these impacts, ranchers have created different strategies. A statistical report was led in Una region for the sorts of rural product and yield pattern honed. The data were gathered using a questionnaire survey was examined using SPSS Editor. These variables are related essentially with surviving components embraced or ranchers throughout the drought time frame. The obtained data were investigated by means of SPSS Editor to acquire frequency and statistical analysis.

Study Area

The Una district lies in the south-western end between Himachal Pradesh and Punjab. The surrounding boundaries are Hamirpur locale in the east, Kangra region in the north and north-east, Punjab State in west and south, Bilaspur in the south-east. The area is extended between 31° 17°52° to 31° 52°0° north latitudes and 75° 58° 21° to 76° 28°25° east longitudes. The region has a total area of 1,550 Sq. kms. and positions tenth in the State arranged by area. It is situated in the Sub-sticky tropical zone of Shiwalik Himalayas in Himachal Pradesh having the vegetation and habitats mainly comprising of ranging from scrubs to coniferous (pine) forests, agricultural lands, rivers. The climate of the region is temperate to tropical in nature, as the terrain differs from plains to high slopes.

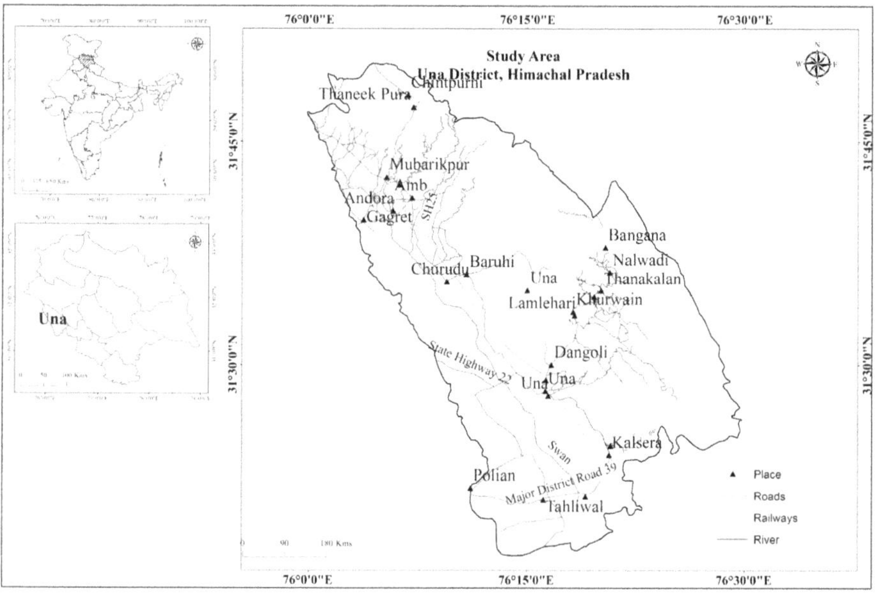

Fig.1: Study Area

Physiography

The total population of area Una is more than three lakhs. The configuration of locale is generally uneven region and falls in the Himalayan foot-slope zone prominently known as Siwalik foot slopes. There are numerous critical slope ranges/Dhars in the area. Prominent among them are Sola Singhi Dhar or Jaswan Dhar, Chaumukhi Dhar, Dhionsar Dhar, Ramgarh Ki Dhar, and Bangar Ki Dhar and all are stretching on the eastern piece of the region. The agro-climatic zone of the Una locale is Western Himalayas, Warm Subhumid Eco-Region. The population of the Swan sub-catchment is around 0.25 million of which 92 % lives in the rustic areas. These Dhars keep running in the locale from north-west to the south-east bearing. Area adjoining Punjab fringe is additionally sloping. Eastern piece of this locale is relatively higher than that of the western part. The elevation of this locale changes between 332 meters and 1,162 meters over the mean ocean level. Soan River is the main river in the region. The development of the banks will mitigate the proportions of a surge in different towns in region Una. Numerous tributaries like Borewalikhad, Barerakhad, Garnikhad, Panjoakhad, Ambwalikhad, Badowalikhad, Humkhad and so forth, join it in the locale from the right and left sides. Soan River and its tributaries shape a valley in the locale known as Soan valley which is flat and generally prolific. It is a regular river and progresses toward becoming itself a tributary of Satluj River outside the locale. In an eastern piece of the region, Lunkharkhad is another khad which flows in the south-easterly bearing and converges with Satluj River. It likewise shapes a restricted valley which is relatively flat and fruitful. The climate of the locale is sub-tropical to temperate and therefore, summers are sweltering and winters are cool. Siwalik ranges encounter substantial rainfall which is for the most part gotten during rainstorm season.

Land Utilize Pattern

The real land utilizes category in the examination area will be the area under fruitless land, which represents 37.81 % of the investigation area. The other dominant land utilize categories are open backwoods (23.77 %), horticultural land (21.24 %) and medium woodland (15.65 %). Geological Area 154.923 ha Geographical area 154.923 ha, Cultivable area 36.879, Forest 18.15 ha, Nonagricultural Land 29.106 ha, Permanent fields and other grazing 13.311 ha, Cultivable wasteland 23.57 ha, Miscellaneous Land 6.728 ha, Barren and uncultivable 23.128 ha, current fallows 1.683ha, other fallows 2.34 ha.

The Agricultural Lands Utilize Pattern	
Net sown area	36.974 ha
Area sowed more than once	34.495 ha
Gross trimmed area	71.469 ha
Net irrigated area	8.556 ha

Source: *Strategic research and extension plan of Una District, Agricultural Technology Management Agency (ATMA), Una District, and Himachal Pradesh.*

Climate

This region is situated at the foot of Siwalik Hills. All through the mid-year climate remains exceptionally hot. In rainy season when rain happens in the upper slopes and water flows down to the plain the Soan River which is a regular river play ruin with the area. Subsequently a significant part of the plain area has been gulped by this river which requires channelization. By channelizing it a significant part of the area will be shielded from soil disintegration and parcel of lives and property can be spared. During winter the days are too cool. During 2009 area got normal yearly rainfall of around 1,329.8 mm. and greatest temperature goes up to 39.50 °C during the long stretch of May and most reduced temperature 2.30 °C is recorded during January. Climate assumes a fundamental job in the field of way of life and economic development of the state, particularly the execution of agriculture, cultivation and tourism sector is firmly related to the execution of rain during the season.

Agriculture

Agriculture is the main occupation of the general population of Himachal Pradesh. It has a vital place in the economy of the state. It gives guide work to 70 for every penny of the total specialists of the state. It additionally is a chief wellspring of state income. Around 17 for each penny of income source originate from agriculture and its united sectors. Likewise, it is additionally the main occupation of the general population in the locale. The agro-climatic conditions prevailing in the locale are ideal for growing yields. The land holdings of the agriculturists in the region are little. The agriculturists develop in excess of two harvests in a year to get most extreme creation from the land. The elevation of this locale differs from 332 meters to 1,162 meters over the mean ocean level. Region got substantial rainfall during rainstorm season. A few endeavors have been made for water harvesting by constructing water storages like irrigation tanks, lakes, check dams and so forth. The division of agriculture is supplying enhanced high yielding assortments of seeds of maize, paddy, wheat, beats, oilseeds, potato, sugarcane, sunflowers and so forth to increase the yield of products. (Sharma, V. (2009).

Materials and Methods

Udmale, P., Ichikawa, Y., Manandhar, S., Ishidaira, H., & Kiem, A. S. (2014) endeavored to understand the country farming network's view of drought impacts on their socio-economic exercises and condition, their adaptation at the family unit level and opinions on government drought mitigation measures. This examination depends on both auxiliary and essential data gathered by means of a survey of 223 farming family units. The outcomes demonstrate that abatement in yield of oats, plant crops, livestock generation and loss of business; all associated with diminished income of agriculturists, and were the most immediate economic impacts of drought. The examination assisted arrangement producers with developing more appropriate drought adaptation approaches in India. Olaleye, O. L. (2010), led a drought managing systems among little scale agriculturists in the Motheo District of South Africa, to determine how ranchers adapt to scarcity impacts through or short of outside impact as far as scarcity alleviation bundles since the legislature. They gathered by administering a semi-organized questionnaire survey to about 200 agriculturists. Questionnaire data were analyzed using SPSS to obtain various analysis such as cross tab, ANOVA, frequency, also calculated relapse analysis. The end is that because of the absence of mindfulness, most ranchers were not set up for drought before the beginning, which made them more defenseless. The validation for the sorts of product utilized and the pattern is finished using the CROPWAT module by considering the different wellsprings of water and yield compose cultivated. Notwithstanding the assessment of drought indices, another goal of the examination is to investigate ranchers coping strategies to drought and their implications to agrarian creation. The CROPWAT display is easy to use and has been effectively used to calculate the impact of climate change on crop water utilize (Teklu, Y., & Hammer, K. 2006). Uncertainties in harvest, soil and weather inputs result in uncertainty in simulated grain yield, evapotranspiration (ET) and nitrogen (N) uptake, which change depending upon the generation condition (Aggarwal, P.

K. 1995). The upside of using the CROPWAT display as a tool for assessing crop water utilize is that it is straightforward and simple to utilize, and its data necessities are less intense than those of other unique models. CROPWAT requires just month to month inputs of climate and rain data, coupled with crop parameters and soils data, to calculate water and irrigation necessities. It isn't a yield development demonstrate however an irrigation planning and management help.

SPSS Editor

SPSS is an exhaustive statistical software bundle that handles all parts of data analysis and data management. SPSS can take data from most sorts of documents and utilize them to generate tabulated reports, graphs, and plots of dissemination and patterns, enlightening statistics, and direct complex statistical investigations. The inherent SPSS Data Editor offers a straightforward spreadsheet-like utility for entering data and browsing the working data record. High-goals, presentation-quality outlines and plots can be created and altered. SPSS enables you to perform numerous examinations on your system that were once conceivable just on substantially bigger machines. SPSS gives a UI that creates a more intuitive condition for analysis for all levels of clients that enable you to perform most errands basically by pointing and clicking. Using the SPSS Viewer, you can handle the yield with greater adaptability. SPSS additionally peruses data documents from an assortment of record formats including Excel, dBASE, Lotus, and SAS.

Choice of the questionnaire in the locale

The Una District spreads wide and conducting questionnaire survey could be a monotonous procedure, subsequently, an example location representing the Una locale was chosen. A drought index analysis was done for the region and the spots were drought is high were understood spatially and transiently and these spots where chosen. The questionnaire was set up to totally understand the cropping pattern, crop type, water resources, climate and other factors anchoring the agrarian development. To accomplish these set points of the investigation, main inclined areas in Haroli, Bangana, Tehsil, Una and Amb were chosen and the choice depended on the way that they experience the ill effects of drought which was dissected with drought indices by remote sensing systems, it ought to likewise be noticed that these areas speak to other networks with comparable attributes which make the findings of the present examination appropriate to such areas. Questionnaire Survey did at these spots on 14.09.2018. Alternate viewpoints of different people with agricultural background in regards with drought were acquired with the utilization of questionnaire survey such as for demography of human, rancher's knowledge about drought, crop type, cropping pattern, water resource, impact, drought impacts for agriculturists and their dependents, and their cop up plan were questioned. The agriculturists were arranged in little and extensive categories using cumulative square root frequency method. The analysis demonstrates that farmers having up to 2.4 hectares of land were little agriculturists while those having over 2.4 hectares were the vast agriculturists. Little-scale agriculturists were fundamentally focused for the survey, and farmers with less than 10 acres were considered. About 50 randomly chose farming individuals.

Analysis and Interpretation

The surveyed information gathered were broke down using SPSS Editor. These variables are associated mainly with the yield compose and rural practices and optional respects with coping instruments embraced or displayed by ranchers during the drought time frame. The collected data was separated and analyzed using SPSS to obtain the frequency test. The outcomes were abridged in light of the original goals of the examination which tried to recognize the horticultural practices, agriculturist's observations, coping instruments, contrasts between coping systems embraced by various family units and in addition the impacts of drought, coping components received through ranchers on their dependent. Questionnaires comprised of human demography, crop type, cropping pattern, and water resources agriculturists' recognition about drought, drought impacts, drought impacts and response on crops, drought consequences for ranchers and their families, and in addition agriculturists' approaches in response to unsettling influences leading to the fluctuations through the drought.

Table 1: Definition of Variable Labels

Was the rainfall profus?	Yes=1; No=2	
Is agro economy is wealth	Yes=1; No=2	
Sources of water?	Open Well. Tank. Ground Water.	Bore well. Tanks Others.
Crops	i)Paddy ii)Wheat iii)Maize	iv)Oil Crops v) Others, specify
Irrigation type	1) Rain Fed 2) Drip irrigation 3) Well irrigation	4) Tank irrigation 5) Others.
Did you change the cropping pattern during the drought period?	Yes=1; No=2	
Did you use analternate source of water for agriculture?	Yes=1; No=2	

Primary Data

The essential data for the present investigation were gathered through a questionnaire survey method. The questionnaires were based open-ended inquiry like yes or no and go between firmly concur and unequivocally oppose this idea. The data on the following viewpoints were gathered, for example, statistic parameters, for example, age, family estimate, structure, sex-ratio, education, occupation, farming background, and so forth cultivate physical inventories like land holding, land utilization pattern, cultivate resources, cultivate actualizes, machinery, and livestock, and so on. Wellspring of water, for example, well water, river, dam, groundwater and its utilization level and so forth., resource utilization including cropping pattern, inputs utilize pattern, human work, composts, and manures, and so forth cultivate creation frameworks including yield of harvests, livestock generation, cultivate costs, creation, utilization, source, for example, cultivate and non-cultivate savvy family unit income.

Table No 2. Frequency analysis

Characteristic Frequency Percentage Rainfall profuse?		
Yes	4	8.0
No	46	92
Total	50	100
Agro Economy is wealthy?		
Yes	48	96.0
No	2	4.0
Total Source of water	50	100
Tank	28	56.0
Open well	19	38.0
Bore Well	3	6.0
Pump set	0	0
Micro Irrigation	0	0
Total Change the crop pattern	50	100
Yes	36	72.0
No	14	28.0

CROPWAT Model

The program utilized for simulating crop yield response to water (CROPWAT) is the best support framework created by the FAO (Aquastat, 2010). Its main capacities are to calculate reference evapotranspiration, trim water prerequisites, and harvest irrigation necessities keeping in mind the end goal to create irrigation plans under different management conditions and plan water supply and to evaluate rain bolstered generation, drought impacts and productivity of irrigation rehearses. It utilizes systems for predicting yields when all the climate, soil and harvest parameters are known. This methodology permits estimation of ETa and Ks, from the ratio of real to potential yield. The test crops are wheat, maize, and paddy. The crop cycle length was 120 days with advancement phases of 15, 30, 40 and 35 days for initiation, vegetative, regenerative and maturity, separately. The general yield coefficient (Kc) values were calculated using Smith, M. (2000), and Van Ranst, E. and Verdoodt, A. (2005) methods. Data on wind speed, temperature, relative moistness, radiation, and rainfall were gathered from the Una meteorological station. Soil concoction properties for the surface soil was determined in the laboratory and data used to calculate the fertility index. The yield factor was calculated at four distinct stages, in particular, the initiation, vegetative, conceptive and maturity stages (Karuku, G. N., Gachene, C. K. K., Karanja, N., Cornelis, W., & Verplancke, H. 2014).

The precipitation data required for CROPWAT 8.0 can be day by day, 10-day time frame (decade) or month to month rainfall, ordinarily accessible from many weather stations. To represent the misfortunes because of runoff or percolation, a decision can be made of one of the four methods given in CROPWAT. As a rule, the productivity of rainfall diminishes with an increase in rainfall and for most rainfall esteems underneath 100 mm/month; the proficiency is approximately 80 %. In the water balance calculations included in the irrigation scheduling some portion of CROPWAT, it is conceivable to evaluate genuine proficiency esteems for various products and soil conditions (Surendran, U., Sushanth, C. M., Mammen, G., & Joseph, E. J. 2015).

To determine the irrigation prerequisites of the investigation area, an assessment was made on the different harvest qualities, for example, length of the development cycle, crop factors, and rooting profundity. Fundamental information gathered from the field included yield and product assortment, planting and harvesting dates. The information on the length of individual development stages, reference evapotranspiration, rooting profundity, suitable exhaustion levels, and yield response factors were gathered from the Research Station.

Crop Evapotranspiration (ETc)

ETo is multiplied by an empirical crop coefficient (Kc) to produce an estimate of crop evapotranspiration (ETc),

$$ETc = Kc \times ETo$$

Calculation of the crop water requirement was completed by re-calling progressively the appropriate climate and rainfall datasets, together with the product documents and the corresponding planting dates entered initially in the model. Soil data was likewise required, together with soil categories, for example, surface, profundity and drainage characteristics and suitability for upland yields. The Soil module is basically data input, requiring the following general soil data; Total Available Water (TAW), greatest infiltration rate, most extreme rooting profundity, and initial soil dampness consumption (Surendran, U., Sushanth, C. M., Mammen, G., & Joseph, E. J. 2015).

Results and Recommendation

Questionnaire information on education level demonstrate that 26.5% of the agriculturists questioned had no basic level of education, whereas about of 73.5% had the formal education which ranged from essential to college instruction. In light of the level of talked with ranchers with no formal education, this infers numerous agriculturists probably won't have the capacity to peruse or compose, which will in general point of confinement their entrance to required data amid dry spell, such circumstance requires a comprehensive planning from assigned government and non-administrative associations to realize mindfulness about any looming catastrophe. The rainfall happens during the South West rainstorm in the long stretch of June – September, and Northeast Monsoon in the period of October – December. The dirt kind reaches from loamy-skeletal soils moderate to serious disintegration which prominently associated with Loamy soils subjected to moderate disintegration.

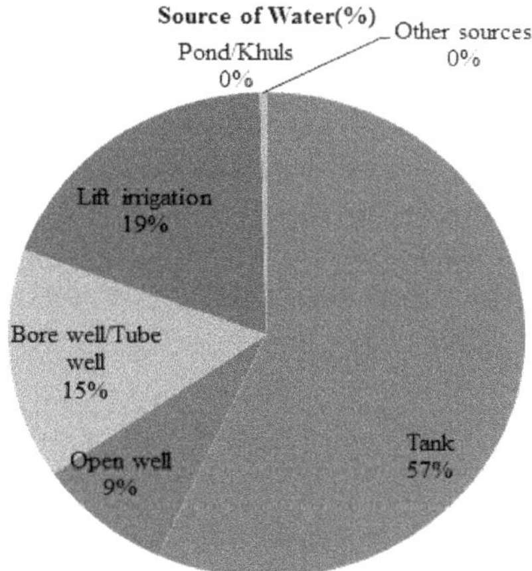

Fig. 2. Sources of Water

The significant sources of water are Tank, Open well, and Bore well/Tube well and pump sets, Micro Irrigation rehearses are not accessible or not mindful in this district and it is gradually evolving. The groundwater accessibility for each sight remove is not as much as the seventy percent of the total areas groundwater level. The significant field crops which create an economy are Wheat, Maize at a greater level taken after by the Rice Oilseeds Fodder at a medium level and at a minimal level Millets, Pulses, and Sugarcanes are cultivated. The vegetables cultivated significantly here is the potato. Mango is the main plant crops taken after by little level yields like Pear, Lime, Galgal, Guava, Litchi, Aonla. The Livestock breeds are as per the following Cattle at a significant level, Buffalo, Sheep, Goats, Pigs, rabbits. The harvest sowing pattern came about as Maize is sowed Kharif-Rain fed 3rd seven day of June to 1st seven day of July, with irrigated Kharif, Maize is sowed 2nd seven day of May to 3rd seven-day of June and paddy is sowed second seven-day of May - first seven-day of July. In both the irrigation sort of the rabi season, wheat is sowed 2nd seven-day of October to 3rd seven-day stretch of December and 1st seven day of November to 4th seven day of December.

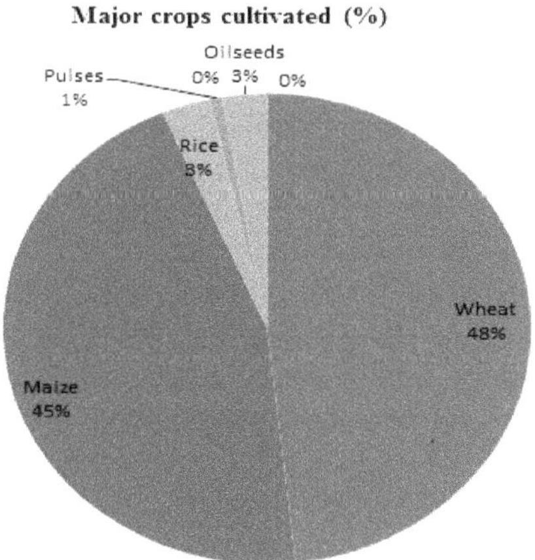

Fig. 3: Major Crops Cultivated (%)

The cropping pattern was dominated by wheat trim that occupied 31,417 ha which was 43.06 for each penny of the total trimmed area (72,955 ha), trailed by maize (30,672 ha) and paddy (1, 976 ha). The offer of money harvests and oilseeds was 2.16 for each penny of the total edited area. Tomato, onion, cultivate pea and cucurbits were the significant vegetable products developed in the region. Maize - wheat was the significant yield rotation taken after by agriculturists which occupied around 80 for every penny of the total area under various harvest rotations. Triple/different cropping framework was likewise followed in irrigated and valley areas, for the most part, falling in Haroli, Bangana Tehsil, Una, and Amb areas.

Results from the SPSS editor were used for the CROPWAT model to predict water use in rainfed agriculture and the irrigated requirement for wheat, paddy, and maize. FAO is used for crop details as a standard measure for the crops considered. In the study area due to perennial crops prevail, constant Crop coefficient values Kc value is considered for all the months in a year which was taken from the table. In the following order of wheat, paddy and maize, for a growing period of (120-150) days (450-650) mm / period of crop water, for a growing period of (90-150) days (450-700) mm / period of crop water and for a growing period of (125-180) days (500-800) mm / period of crop water is required. The model predicted deficiency as (15- 32) % in the crop water requirement for the crops considering the various climate, water sources, and the crop evapotranspiration. The crop evapotranspiration (ETc) for wheat were predicted as 435.6 mm while needed evapotranspiration (ETa) was 561.5 mm for the total period time. Similarly, the crop evapotranspiration (ETc) for maize were predicted as 405.6 mm while needed evapotranspiration (ETa) was 436.6 mm for the total period time. The suggestion was that as an addition of (20-25) % of irrigation water has to be added to overcome agricultural deficiency.

As we can see, the efficiency of the agricultural practices during the drought are less than actual need with respect to the model's result, hence a prominent coping up strategy farmer must follow to run the agribusiness on track without any huge loss. As we all know, drought cannot be predicted, hence farmers must be prepared at all times and ready to adapt to the strategies. The adapting up measures they should take after varies as Change in crop types/cropping design, Crop administration, Soil supplement, and soddenness protection measures, Wheat grazing crop by cows, Inter-development (soil mulching), Conservation Furrow, Foliar administration, Puckering in maize, don't use engineered inventions for weed administration under strain etc. Although as a moderation measure agriculturists used diverse customary practices as dry spell preparation and adjustment measures, less tendency was given to their adaption. It is found through survey that low level of education, little land property size and low livelihoods were genuine imperatives in apportionment of these adjustment methodologies. Outstanding consideration should be set to these impediments government and other private areas while conspiring and planning dry season methodologies for expanding system solidarity to future dry spell events. Other than nuclear family level adjustment measures, regulatory procedures accepted to a great degree fundamental employment in adjusting to dry season. As a reaction to certifiable dry season events in the express, the organization has endeavored distinctive help measures. It was seen that the relief measures offered mitigation to affected nuclear families somewhat, however for a more noteworthy great long haul arranging must be completed thinking about the limitations.

References

Aggarwal, P. K. (1995). Uncertainties in crop, soil and weather inputs used in growth models: implications for simulated outputs and their applications. *Agricultural Systems, 48*(3), 361-384.

Aquastat (2010). Information system on water and agriculture. FAO, http://www.fao.org/nr/water/aquastat/main/index.stm

Karuku, G. N., Gachene, C. K. K., Karanja, N., Cornelis, W., &Verplancke, H. (2014). Use of CROPWAT model to predict water use in irrigated tomato production at Kabete, Kenya. *East African Agricultural and Forestry Journal*, 80(3), 175-183.

Olaleye, O. L. (2010). *Drought coping mechanisms: a case study of small scale farmers in Motheo district of the Free State province* (Doctoral dissertation).

Sharma, V. (2009). *A Study in Maize Production, Pattern of Residues Use and their Marketing in Una District of Himachal Pradesh* (Doctoral dissertation, CSKHPKV Palampur).

Singh, D. (2015). Avifaunal Diversity of Hamirpur District, Himachal Pradesh, India. *International Journal*, 3(12), 265-276.

Smith, M. (1992). CROPWAT: *A computer program for irrigation planning and management* (No. 46). Food & Agriculture Organization

Surendran, U., Sushanth, C. M., Mammen, G., & Joseph, E. J. (2015). Modelling the crop water requirement using FAO-CROPWAT and assessment of water resources for sustainable water resource management: A case study in Palakkad district of humid tropical Kerala, India. *Aquatic Procedia*, 4, 1211-1219.

Teklu, Y., & Hammer, K. (2006). Farmers' perception and genetic erosion of tetraploid wheats landraces in Ethiopia. *Genetic Resources and Crop Evolution*, 53(6), 1099-1113.

Udmale, P., Ichikawa, Y., Manandhar, S., Ishidaira, H., & Kiem, A. S. (2014). Farmers perception of drought impacts, local adaptation and administrative mitigation measures in Maharashtra State, India. *International Journal of Disaster Risk Reduction*, 10, 250-269.

Verdoodt, A., & Van Ranst, E. (2006). Environmental assessment tools for multi-scale land resources information systems: a case study of Rwanda. *Agriculture, Ecosystems & Environment*, 114(2-4), 170-184.

Vishwa, B., Chandel, S., & Karanjot, K. (2013). Drought in Himachal Pradesh, India: A Historical-geographical perspective, 1901-2009. *Transactions*, 35, 260-273.

7

Resource Use Efficiency and Comparative Costs: A Case of Farming System in Karnataka

Vijayachandra Reddy. S. and S. M. Mundimani

Introduction

India is traditionally recognized as country of organic agriculture for good quality produce cultivation practices, in spite of this application and adoption of modern scientific and high input use in agriculture has placed organic farming in last position of acceptance for production of agricultural produce. However, the escalating consciousness of quality and better food safety standards for long term goals especially in rain-fed areas observed as resource deficit farming, in these areas organic farming came has boon as source of alternative farming providing good source of income for their quality produce apart improving the standard of living by generating high income levels when compared to other farming practices (Murmu, 2018).

In addition to increasing area of operations from last few years, it also aids in achieving food security plans through institutional mechanism by administration support during 12th FYP and also in the 11th FYP of Horticulture and Plantation Working Group of planning commission. To stabilize the faith among farming community, government efforts to make arrangement to provide farmers produce to link with market led support to earn better prices. During the year 2007, Indian Government has evidently mentioned about development of organic farming in different parts of India. Though institutional arrangements are coming forward in this line, the era of organic farming suffers from inadequate institutional support especially in areas such as extension, research and maintaining quality standards for certifications at international markets which needs more introspective monitoring of management practices. All these initiatives will certainly boost the World Trade Regime (WTO) through export oriented programmes for agriculture and horticulture produce. It was interesting to notice that, during the study period, If farmer were are asked about organic farming most of the farmers revealed that "they do not use pesticides, fertilizers or any chemicals for plant protection and they cultivated crops by means of natural ways" (Zanoli, 2004).

Thus, it establishes roots of natural biological process linking to sustainable methods for management of resources in ecosystem. This process enhances the adoption of more natural ways of resource conservation rather than exploiting it (McEwen and Milligan, 1991). It creates a favorable sphere for biological resources which strengths the soil and water holding capacity of soil and makes the soil more fertile in short duration and reduces the impact of environmental degradation and encourages agriculture ecosystem. The process of organic agriculture not only reduces harmful effects on soil and water use, it also finds a way to propel the natural resource base for overall development on sustainable basis especially in organic farming. Hence the traditional farming practices have been converted into modern methods such as intensive with more alternative methods such as Low Input method, minimum tillage, no tillage, crop rotation and inter cropping with legumes. This change of pattern not only influences cost of production and also quality of produce.

Among Sustainable development dimensions such as Economic, Social and Environmental, Considering the Economic aspects as important dimensions of sustainability, Organic farming which supports for sustainability in long run and also helps in tackling the problem of food security when compared to conventional farming systems. Apart from sustainability point of view, the organic produce have good market value in terms of quality and health benefits both at regional and international markets with great value addition in process value chain of agricultural commodities. The probable reason behind growing increasing demand for organic produce is due to increase care of health habits and also shifts in dietary habits followed by different parts of many developed countries.

In era of agriculture, Organic farming is an eco-friendly and most effective way to assess the overall sustainability of farming practices. It plays a vital role in managing the local resources which are ecosystem based management by the farming community leads to development of local farm resources practices which are contributing towards sustainability.

Objectives and Methodology

The objectives of the paper are (i) to analyze the resource use efficiency, and (ii) to evaluate the comparative costs of the organic farming systems in Bijapur District of Karnataka State.

Production function analysis: In order to assess the resource use efficiency of inputs used by sample farmers in different organic farming systems, the Cobb-Douglas production function was estimated for different organic farming systems using the following form of the function.

$$Y = aX_1^{b1} X_2^{b2} X_3^{b3} X_4^{b4} X_5^{b5} X_6^{b6} X_7^{b7} e^u$$

Where,

Y = Gross income of the farm (Rs./ha)

X_1 = Expenditure on Seeds (Rs./ha)
X_2 = Expenditure on farm yard manure (Rs./ha)
X_3 = Expenditure on biofertilizers (Rs./ha)
X_4 = Expenditure on organic manures (Rs./ha)
X_5 = Expenditure on bio pesticides (Rs./ha)
X_6 = Expenditure on human labour (Rs./ha)
X_7 = Expenditure on bullock labour and machine labour cost (Rs./ha)
a = Constant
b_i = Production elasticities, i = 1, 2,...6
U = Random error

All variables were measured in monetary values. The above function was converted into the linear form through logarithmic transformation and written as

Log Y = log a + $b_1 \log X_1 + b_2 \log X_2 + b_3 \log X_3 + b_4 \log X_4 + b_5 \log X_5 + b_6 \log X_6 + b_7 \log X_7 + \log u$.

Allocative efficiency: The ratio of the MVP's to MFC's of individual resources was used to judge the allocative efficiencies. The computed Marginal Value Product (MVP) was compared with the Marginal Factor Cost (MFC) or opportunity cost of the resource to draw inferences. A resource is said to be optimally allocated when its MVP = MFC. The marginal value product for each input was calculated at the geometric mean levels of the respective resources by using the formula,

$$\text{Marginal Value Product (MVP) of } X_i = bi \frac{Y}{Xi}$$

Where,
Y = Geometric mean of gross income
X_i = Geometric mean of ith resource
b_i = Production elasticity of ith resource

The Marginal Value Product (MVP) was compared with the Marginal Factor Cost (MFC) to arrive at optimal use of the resources. This analysis was carried out in order to identify the possibilities of increasing gross returns under a given farm situation.

Dairy Enterprise: Production function technique was used to assess the resource use efficiency in dairy enterprise namely dry fodder, green fodder, concentrates, human labour, veterinary expenditure etc. on income. Cobb-douglas type of production function of the following form was fitted to the data.

$$Y = a\, X_1^{b1}\, X_2^{b2}\, X_3^{b3}\, X_4^{b4}\, X_5^{b5} \ldots\ldots\ldots\ldots X_n^{bn}.\, e^U$$

Where,
Y = Total income (Rs/animal)
X_1 = Green fodder (Rs/animal)
X_2 = Dry fodder (Rs/animal)
X_3 = Concentrates (Rs/animal)
X_4 = Veterinary expenditure (Rs/animal)
X_5 = Human labour (Rs/animal)
a = Intercept
b1, b2, b3 .... bn = Regression coefficients or production elasticities
U = Random error

Goat Enterprise: Production function technique was used to examine the effect of different factors namely, green fodder, concentrates, veterinary expenditure, human labour etc. on income. Cobb-douglas type of production function of the following form was fitted to the data.

$Y = a\, X_1^{b1}\, X_2^{b2}\, X_3^{b3}\, X_4^{b4}\, X_5^{b5} \ldots\ldots\ldots\ldots X_n^{bn}.\, e^U$

Where,

Y = Income (Rs. /animal)

X_1 = Green fodder (Rs. /animal)

X_2 = Concentrates (Rs. /animal)

X_3 = Veterinary expenses (Rs. /animal)

X_4 = Human labour (Rs. /animal)

a = Intercept

b1, b2, b3 .... bn = Regression coefficient or production elasticities

U = Random error

These forms of the Cobb-douglas type of production functions were used in general for assessing resource use efficiency of crops as well as other enterprises followed in major farming systems. However, only those inputs were used for particular enterprises were retained in particular function. Inputs which were not used by majority of farmers in particular enterprise were removed from the production function model fitted to the data.

Results and Discussion

A. Resource use Efficiency in Bijapur District

The estimated production coefficients and MVP to MFC ratios for various resources used under organic farming systems in Bijapur district were computed and the results are presented in Table 1. The findings of production function for organic farming systems FS-I in Bijapur district shows that, the regression coefficients of all the resources were positive except for green fodder (-3.83), bullock labour (-1.70), organic manure (-0.39) and machine labour (-0.08), The coefficients were positive for seed, human labour and bio pesticide and significant at one per cent level, where as cost of concentrates and dry fodder were statistically significant at 5 per cent level and all other resources coefficients were non-significant. The variables incorporated in the production function explained 97.00 per cent variation in the dependent variable as indicated by value of coefficient of multiple determination (R) was 0.97.

The ratios of MVP to MFC were less than unity for bullock labour (-58.07), machine labour (-1.28), organic manure (-19.17) and green fodder (-37.13). For other variables such as seed (139.40), farmyard manure (41.73), human labour (1.27), bio pesticide (121.56), dry fodder (189.25) and concentrates (44.97) the ratios were greater than unity. The summation of regression coefficients indicated increasing returns to scale (1.08).

The resource use efficiency of FS-I in Bijapur district is presented in Table 1. The coefficient of seed (2.09) was found to be significant at one per cent showing that for one rupee increase in cost of seed the gross returns increase by Rs. 2.09. Similarly, human labour (0.37), bio pesticide (0.67), dry fodder (6.61) and concentrates (1.92) found significant at one per cent level. For one per cent increase in the cost of human labour, bio pesticide, dry fodder and concentrates the gross returns would increase by Rs. 0.37, Rs.0.67, Rs. 6.61 and Rs. 1.92 respectively.

The ratios of MVP to MFC were less than unity for bullock labour, machine labour, organic manure and green fodder. It indicates that these resources were over utilised hence farmer has to stop the use of these resources. For other variables such as seed, farmyard manure, human labour, bio pesticide, dry fodder and concentrates the ratios were greater than unity indicating that these resources were underutilised still there is scope to increase the use of available resources (Hildermann et al., 2010; Clabby, 2010). Since the returns scale was more than one, the gross returns will increase by more than one rupee.

Organic FS-II

In case of farming systems II, the regression coefficient of all the resources were positive except for seed (-0.49), organic manure (-0.03) and green fodder (-0.07). The coefficients of farmyard manure, machine labour and bio pesticide were statistically significant at one per cent level and coefficients of all other resources were non-significant (Nagaraj, T., Khan, H. S. S. and Karnool, N. N., 1998). It would be observed that the fitted production function found to be good fit to the data as coefficient of multiple determination (R) was 0.92. The ratios of MVP to MFC were less than unity for seed (-9.30), organic manure (-0.55) and green fodder (-2.40). For other variables such as farm yard manure, human labour, bullock labour, machine labour, bio pesticide and dry fodder were greater than unity. The summation of regression coefficients indicated increasing returns to scale (1.36).

It would be observed that the fitted production function found to be good fit to the data as coefficient of multiple determination (R) was 0.92. The ratios of MVP to MFC were less than unity for seed, organic manure and green fodder indicating that these resources were over utilised. For other variables such as farmyard manure, human labour, bullock labour, machine labour, bio pesticide and dry fodder were greater than unity indicating these resources were underutilized (Bhende *et al.*, 2007). The summation of regression coefficients indicated increasing returns to scale.

Organic FS-III

The production co-efficient for farming systems III in Bijapur district revealed that the coefficients found positive and failed to exert significant impact on production were seed, human labour, bullock labour, bio pesticide and green fodder. The dry fodder found to have positive and significant impact on production, while machine labour found to have negative and significant impact on production (Kumara Charyulu and Subho Biswas, 2010). The coefficient of multiple determination (R) was 0.83 indicating nearly 83.00 per cent of variation in dependent variable was explained by the independent variables considered for the study.

The MVP to MFC ratios were more than unity for seed followed by human labour, bullock labour, bio pesticide, dry fodder and green fodder. For other variables such as farmyard manure, machine labour, organic manure and concentrates were lesser than unity. The summation of regression coefficients indicated increasing returns to scale (1.12).

Table 1: Resource use efficiency under different organic farming systems in Bijapur district

Sl. No.	Particulars	Parameters	Bijapur district					
			Farming system-I		Farming system-II		Farming system-III	
			Production Coefficients	MVP:MFC Ratios	Production Coefficients	MVP:MFC Ratios	Production Coefficients	MVP:MFC Ratios
1	Intercept	a	34.60 (20.37)		0.57 (1.88)		89.54 (48.05)	
2	Seed	b 1	1.79** (1.03)	139.40	-0.49 (0.35)	-9.30	3.81 (4.24)	68.04
3	FYM	b 2	1.12 (0.67)	41.73	0.50** (0.11)	15.86	-1.53 (1.02)	-42.46
4	Human Labor	b 3	0.37** (0.54)	1.27	0.20 (0.16)	0.95	0.20 (0.78)	0.64
5	Bullock Labor	b 4	-1.70 (0.82)	-58.07	0.055 (0.11)	2.16	1.75 (1.75)	107.48
6	Machine Labor	b 5	-0.08 (0.10)	-1.28	0.41** (0.15)	9.85	-6.37** (1.78)	-101.07
7	Biopesticide	b 6	0.67** (0.38)	121.56	0.66** (0.13)	36.79	1.92 (1.18)	145.57
8	Organic manure	b 7	-0.39 (0.43)	-19.17	-0.03 (0.08)	-0.55	-2.09 (2.44)	-31.25

9	Dry fodder	b 8	1.61* (0.53)	189.25	0.32 (0.16)	32.46	2.67** (1.21)	125.28
10	Green fodder	b 9	-3.83 (3.73)	-37.13	-0.07 (0.10)	-2.40	1.68 (6.15)	42.41
11	Concentrates	b 10	1.52* (1.44)	44.97	-		-0.92 (2.45)	-6.89
12	R^2		0.97		0.92		0.83	
13	F- value		238.73		581.17		92.39	
14	Returns to scale		1.08		1.36		1.12	

Note : Figures in parentheses indicates the respective Standard error **and * Denote significance level at 1 % and 5% respectively

The production co-efficients for FS-III in Bijapur district revealed that the coefficients found positive and failed to exert significant impact on production were seed, human labour, bullock labour, bio pesticide and green fodder. The dry fodder found to have positive and significant impact on production, while machine labour found to have negative and significant impact on production. The MVP to MFC ratios were more than unity for seed followed by human labour, bullock labour, bio pesticide, dry fodder and green fodder showing that these resources were underutilised. For other variables such as farmyard manure, machine labour, organic manure and concentrates were lesser than unity indicating these resources were over utilised. The summation of regression coefficients indicated increasing returns to scale.

B. Comparative Economics of Major Organic Farming Systems in the Study Area

The comparative economics of major organic farming systems in the study area are presented in Table 2. The organic farming systems were selected based on the percentage of adoption. The organic farming systems in Bijapur district reveals that, the total variable cost was maximum in farming systems II (Rs. 80187.29) as compared to farming systems III (Rs. 78079.99) and farming systems I (Rs. 60300.56). Similarly, fixed cost was highest in FSII (Rs. 29687.68) followed by farming systems III (Rs. 16470.16) and farming systems I (Rs. 15739.14). The total cost was found to be more in farming systems II (Rs. 109874.97) as compared to farming systems III (Rs. 94550.15) and farming systems I (Rs. 76039.70). The gross returns was maximum in farming systems II (Rs. 143634.79) followed by farming systems III (Rs. 118181.40) and farming systems I (Rs. 108542.63). Similarly, net returns was highest in case of farming systems II (Rs. 33759.33) as compared to farming systems I (Rs. 32502.92) and farming systems III (Rs. 20614.27). The benefit cost ratio was highest in farming systems II (1.44) followed by farming systems I (1.41) and farming systems (1.25) III (Inder pal and Grover, 2011).

The organic farming systems were selected based on the percentage of adoption. The organic farming systems in Bijapur district revealed that (Table 2), the total variable cost was maximum in FS-II (Maize + Chickpea + Lime + Goat rearing) as compared to FS-III (Groundnut + Wheat + Dairy) and FS-I (Greengram + Sorghum + Dairy). Similarly, the gross returns was maximum in FS-II (Maize + Chickpea + Lime + Goat rearing) because of more returns from lime cultivation and goat rearing. Similarly, net returns available with farmer after meeting the expenses incurred in production was highest in case of FS-II (Maize + Chickpea + Lime + Goat rearing) as compared to FS-I (Greengram + Sorghum + Dairy) and FS-III (Groundnut + Wheat + Dairy).

Table 2: Comparative Economics of Major Organic Farming Systems in the Study Area (Rs/Farm)

Sl. No	Particulars	Bijapur*		
	Costs	**FS I**	**FS II**	**FSIII**
1	Total variable cost	60300.56	69956.25	78079.99
2	Total fixed cost	15739.14	29687.68	16470.16

Sl. No	Particulars	Bijapur*		
3	Total cost	76039.70	99643.93	94550.15
	Returns			
1	Gross returns	108542.63	143634.79	118181.40
2	Net returns	32502.92	43990.37	20614.27
3	B:C ratio	1.41	1.44	1.25

***In Bijapur District**

FS-I Green gram+Sorghum+Dairy

FS-II Maize+ Chickpea+Lime+Goat rearing

FS-III Groundnut+Wheat+Dairy

References

A. C. (1991). Agriculture Canada, *Progress Report on the Development of Organic Certification Standards and Accreditation System*, December, p. 1.

Bhende, M. J. and Kalirajan, K. P., (2007). Technical efficiency of major food and cash crops in Karnataka (India). *Indian Journal of Agricultural Economics* 62(2), 176-190.

Clabby, C. (2010). Does Peak Phosphorus Loom? Scientists make the case that easily accessible supplies of an essential element are being depleted, *American Scientist*, 98, 291-292.

McEwen, F. L. and Milligan, L.P. (1991). *An Analysis of the Canadian Research and Envelopment System for Agriculture/Food*, Report commissioned by the Science Council, July, p. 20.

Hildermann, I., Messmer, M., Dubois, D., Boller, T., Wiemken, A. and Mader, P. (2010). Nutrient use efficiency and arbuscular mycorrhizal root colonisation of winter wheat cultivars in different farming systems of the DOK long-term trial, *Journal of the Science of Food and Agriculture*, 90, 2027-2038.

Singh, I. P. and Grover, D. K. (2011). Economic Viability of Organic Farming: An Empirical Experience of Wheat Cultivation in Punjab. *Agricultural Economics Research Review*, 24, 275-281.

Murmu K (2018). Organic Farming Stewardship for Sustainable Agriculture. *Agricultural Research and Technology: Open Access Journal.* 13(3), 555883

Kumara, C. D and Biswas, S. (2010). Economics and Efficiency of Organic Farming vis-à-vis Conventional Farming in India, *Working Paper No. 2010-04-03*, April 2010, Indian Institute of Management Ahmadabad (IIMA), INDIA.

Nagaraj, T., Khan, H. S. S. and Karnool, N. N., (1998). Resource-use efficiency in various crops under different cropping systems in Tungabhadra Project Command Area. *Agricultural Situation India, 55*(3), 135-139.

Zanoli, R. (2004). *The European Consumer and Organic Food*, Aberystwyth School of Management and Business, University of Wales.

8

Sustainable Agricultural Development in Punjab

Harpreet Kaur

Introduction

Agriculture the supplier of that basic human need, nutrition is the world's largest user of land, occupying more than One third of Earth's terrestrial surface and also using vast amount of water (UN, 2013). Agriculture affects our daily life in several ways both directly as well as indirectly. Humans expect agriculture to supply sufficient nutrients, economically and culturally valued foods fibres and other products. The Development of agriculture sector was essential for the rise and survival of early civilizations in any society. In the same way Indian economy remains a predominantly agrarian economy and agriculture is demographically the broadest economic sector of the nation, with 58 per cent of India's population dependent on the sector for their livelihoods. The agriculture sector in India has undergone significant structural changes since independence. In the same way Punjab's agricultural sector comprised of almost 20 per cent of the total workforce of the state in 1970-71, which has reduced marginally to 19 per cent in 2000-01. State produces 61 per cent of India's wheat and 30 per cent of India's national pool but by the peak of the green revolution in the 1980s, the state contributed 73 per cent of the wheat and 45 per cent of rice to the national pool.

Punjab has been a star performer in agriculture during heydays of Green Revolution. The rate of progress was quite substantial in the initial years of green revolution mainly due to exponential increase in agricultural productivity and expansion in gross cropped area. It was stellar performance of Punjab, first witnessed in large wheat surpluses and then similar rise in production of rice, which helped India free itself from the PL 480 food aid and its associated political strings. Punjab became a symbol of India's grain surpluses, giving India the much needed food security. But after 1985-86, green revolution started greying and growth in Punjab agriculture slowed down to 3 percent per annum over the period (Gulati, Roy, & Hussain, July 2017). However, in later years, the progress in agricultural production has slowed down and signs of stagnation are quite visible. Decelerating agri-growth, Punjab has lost its pre-eminent position of being the state with highest per capita income in India. There have been some disturbing developments in Punjab agriculture during the post Green Revolution period. The Green Revolution becomes a paradoxical process. The father of India's Green Revolution, Dr M.S. Swaminathan while talking to Economic Times Kolkata Bureau in 2004 had said that our 'Green Revolution' has been abused into 'Greed Revolution' (Sidhu & Rangi, 2004).

According to Vandana Shiva, if on the one hand, it offered technology as a substitute to both nature and politics in the creation of abundance and peace. On the other hand, the technology itself demanded more intensive natural resources use along with intensive external inputs. As a result, new relationships between the state and cultivators, between international interest and local communities within the agrarian society developed. The Green Revolution ended India's dependence on other countries. But it is also true that Punjab agriculture is now facing serious problems. There is no scope for further expansion of the area for cultivation or increase in cropping intensity. The increase in prices of inputs and labour has pushed the cost of production many times. That raises fundamental question, how can it get back to high growth path. Under this scenario the significant strategies of sustainable agriculture has been emphasize and justified in the present paper for upcoming generations of Punjab.

Data Sources

The data used to study the importance of sustainable development is based on the secondary data purely. The journal papers, article, web links, books and government reports have been used as source of information. There is no empirical touch to this paper.

Discussion

Recent phenomenon of converting large-scale forest areas, grazing lands and waste lands into croplands to support the rising population has caused ecological imbalance and atmospheric pollution. With no further scope for expansion of agricultural land efforts have been made to enhance the production of food grains using high-yielding variety of seeds, fertilizers and irrigation along with advanced farm equipments. No doubt green revolution ended India's dependence on other countries. But it is also true that agriculture sector is now facing serious problems. Discussion of these problems replicates the importance of sustainable agriculture practices.

Wheat, which was traditionally both a food and a cash crop, has been joined by rice as became commercial crop for the Punjabi farmer at the expense of the traditional commercial crops such as ground nut, rapeseed and mustard (Sidhu and Sidhu 1988). Many traditional and most conventional farm practices are not ecologically sustainable: they overuse natural resources, reducing soil fertility, causing soil erosion, and contributing to global climatic change. Sustainable agriculture has several major advantages over both traditional and conventional practices.

Agriculture cannot be sustainable unless it is environmentally and economically viable over the long term. Over the time, so many practices and technologies have been developed by researchers in the direction of promoting stability in agricultural production contrary to the assault of seasonal variations. Adoption of such resilient practices and technologies by farmers is not an option for them but is an essential for their smooth survival. Therefore, the technology up gradation is not a sole concern, but this up-gradation must incorporate the climate resilient technologies and practices that is environmental and social safeguards, efficiency in resource-use, sustainability and long-term development of agriculture assume greater importance. Such an approach can ensure adaptation gains and instant benefits to farmers along with possible reduction in GHG emissions and global warming potential of agriculture and leads to Climate Smart Agriculture (NICRA, 2014).

"Climate Smart Agriculture aims to sustainably increase productivity and income, build resilience to climate change, reduce greenhouse gas emissions and enhance achievement of national food security and development goals" (CCAFS, 2013). In this context, some of the Climate Smart and Smart practices are pointed out as follows:

Water-smart technologies: A major impact of climate change on agriculture will be experienced in the form of water stress. Therefore, land development, water conservation and harvesting measures in both rain-fed and irrigated agriculture, crop planting, and irrigating techniques, among others, could form a package of practices to help overcome the climate-imposed stress (Sharma et al. 2006). Interventions that reduce water requirements to produce the same or a higher level of yield called water smart technologies. Specific technologies in this group includes the planting of crops that use less water and rainwater management, better irrigation techniques such as rainwater harvesting, micro-irrigation techniques and drip irrigation, laser land levelling and system of rice intensification etc.

Energy-smart technologies: Energy is the most important component for present-day agriculture. Although energy contributes to greenhouse gas emissions, the lack of it is a major barrier to agricultural development. Conservation agriculture is an energy-smart emerging intervention for sustainability (Lumpkin and Sayre 2009). Energy Smart technologies are technologies that help reduce energy consumption during land preparation without affecting yield levels. These also help reduce water requirements for crops. For example direct seeded rice, zero/minimum tillage etc. (Taneja et. al, 2014).

Nutrient Smart Technologies: Nutrient Smart Technologies supplements or avoids chemical fertilizer usage for harvests and augment nutrients in the soil. It incorporates green manure, integrated nutrient management and leaf color chart etc. Agricultural practices that make sure soil and nutrient management, carbon management and nitrogen management can contribute to the conservation of resources, sequestration of carbon, and preservation of future food security (Lal et al. 2011).

Introducing Tolerant Crops: Drought is a major climatic aberration that is confronted in most of the states and therefore, drought-tolerant crop varieties should be used as a smart technology for adaptation according to given climatic condition. Drought tolerant variety refers to seed variety that is tolerant to drought or relatively dry weather conditions (Taneja et. al, 2014). For example, Pearl millet and sorghum are the most inherently drought-tolerant of all the major staples and main cereal grain harvests in the dry lands, providing food, feed and, in the case of millet, fuel and construction material as well (CGIAR, 2017).

Organic Farming: In the 20th century as a response of rapidly changing farming practices and intensive increased use of synthetic substances, organic farming is emerged as an alternative agricultural system. In this system organic criterions are aimed to permit the use of natural substances like fertilizers of organic origin such as green manure, compost, and bone meal while prohibiting or strictly limiting synthetic substances. In this system crop rotation and companion planting, biological pest control, mixed cropping etc. techniques are encouraged. Organic farming is not only eco-friendly but it is also highly remunerative.

"Lakshmana Sharma, a castor farmer of Ayavaripalle of Midgilmandal in Mahabubnagar district, reaped six to eight tons of the crop compared four tons by others by spending only Rs 500 and applying organic fertilizers to castor crop raised in his two-and-half acre farm. He is hopeful of reaping at least 20 quintals of castor seeds at the end of the season", (The Hans India, 2017). "Uttarakhand has wisely declared several districts organic, which mean that the farmers must undertake only organic farming. The good news is, it isn't just any old government order. The farmers, it seems, agree with this order entirely," (Hindustan Times, 2017). Punjab Government should also take such decisions for suitable areas.

Weather Smart Decisions: (Crop Insurance, Weather Advisories) Weather Smart decisions may include the awareness of farmers for getting information about weather conditions and financial security in terms of crop insurance.

Crop insurance has been a major initiative to help farmers deal with the vagaries of weather. Two schemes, the Weather Based Crop Insurance Scheme and the National Agricultural Insurance Scheme, are in operation. The first scheme compensates farmers for financial losses attributable to incidences of adverse weather conditions such as rainfall, temperature, frost, and humidity. The National Agricultural Insurance Scheme provides financial support to farmers in the event of crop failure due to natural calamities, pests, or diseases (Taneja et al, 2014).

Information and Technology: Access to information and improved communication is a crucial requirement for sustainable agricultural development. Information and Communication Technologies, can play a vital role by establishing a network among the researchers, extension personnel, and farmers for transfer of information. Consequently, technology penetration has not only widened the scope of agriculture, but brought government's attention to higher production vis-a- vis empowerment of rural farmers while providing agricultural information at their door step. Effective systems have been developed followed by major interventions like Green Revolution of the late 1960's and Information and Communication Technology (ICT) in the late 1990's.

Conclusion

It is the need of hour that farmers should wisely adapt to the climate change to maintain farm productivity and income so that agriculture development can sustain for long time. Moreover for protection of livelihoods of small and marginal farmers, there is the supreme significance of enhancing agricultural resilience to climate threat. Adoption of climate smart practices and technologies by farmers is not an option for them but is an essential for their smooth survival. The government should seek to reorient and align policies to promote climate smart agriculture and affect land use, crop choices, fertilizer use, irrigation practices and energy inputs to complement each other towards the common goals of sustainability and growth. Simultaneously, through better regulation, the government must leverage the potential of markets and market mechanisms to promote resource use efficiency in agriculture. The new paradigm must combine sticks with carrots and both with better information.

References

Baweja, S., Aggarwal, R., & Brar, M. (2017). *Groundwater Depletion in Punjab, India.* Encyclopaedia of Soil Science, Third Edition: Three Volume Set .

Gulati, A., Roy, R., & Hussain, S. (2017). *Getting Punjab Agriculture Back on High Growth Path: Sources, Drivers and Policy Lessons*. CRIER.

Sidhu, M.S. & Rangi, P. S. (2004). Dynamics of Unemployment in Punjab : An Analysis. In B. Singh, *Punjab Economy : Challenges and Strategies* (p. 63).

Singh, S., & Tiwar, C. (2012). Challenge of Managing Soil Fertility Depletion: A Case Of Tarai Areas Of Uttarakhand, India. *International Journal of Rural Studies*, 19(2), 1-6

Sharma, B. R., K. V. Rao, K. P. R. Vittal, & Amarasinghe, U. A. (2006). *Realizing the Potential of Rain-Fed Agriculture in India.* Draft prepared for the IWMI-CPWF project on Strategic Commands. Bhubaneswar, India: Directorate of Water Management Research, Indian Council of Agricultural Research.

Lumpkin, T. A., & Sayre, K. (2009). Enhancing Resource Productivity and Efficiency through Conservation Agriculture. *In 4th World Congress on Conservation Agriculture: Innovations for Improving Efficiency Equity and Environment,* p.3-9. New Delhi: 4th World Congress on Conservation Agriculture.

Taneja, G., Pal, B. D., Joshi, P. K., Aggarwal, P. K., & Tyagi, N. K. (2014). Farmers' preferences for climate-smart agriculture: An assessment in the Indo-Gangetic Plain, *IFPRI Discussion Paper 01337*. New Delhi: International Food Policy Research Institute (IFPRI)

CGIAR (2017). Drought-Tolerant Crops for Dry lands. Available at http://www.cgiar.org/web-archives/www-cgiar-org-impact-global-des_fact2-html/ retrieved on February 05, 2017.

The Hans India (2017). Organic cultivation boosts castor crop yield. Available at http://www.thehansindia.com/posts/index/Khammam-Tab/2017-02-27/Organic-cultivation-boosts-castor-crop-yield/283486 retrieved on February 28, 2017

9

Prospect of Sustainable Agricultural in Punjab, India

Harminder Singh and L. T. Sasang Guite

Introduction

The two dimensions in the concept of sustainable (making better and maintain) envisage sustainability on inter-generational equity implying availability of resources in terms of human well-being and opportunities for future generations. The primary focus of sustainability aims at meeting present human needs as well as preserving the environment so that the need could also be fulfil for the future generation. Development plans are design to ensure that: Sustainable and equitable use of resources, preventing further damage to our life support systems; and, conserving and nurturing the bio-diversity gene pool and other resources for long-term food security. Sustainable Development ("sustain" comes from the Latin word "sustinere" sus-from below and tenure- to keep in existence or maintain which implies long-term permanence) is a multidimensional concept with three interacting terms (Fig No.1); ecological security, economic efficiency and social equity. Thus, the uniform relationship of the three terms is utmost importance in achieving sustainable development.

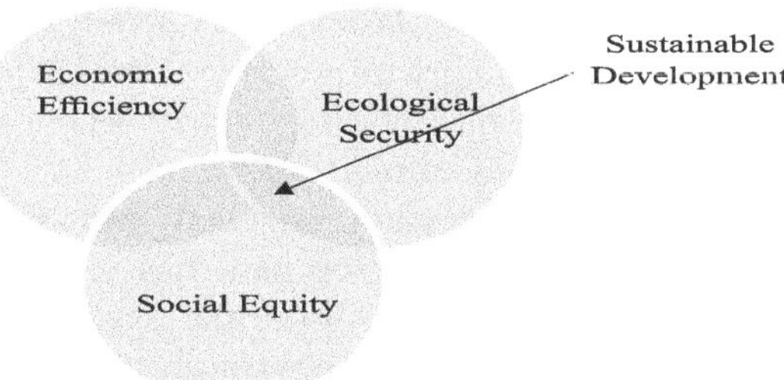

Sustainable agriculture is one such aspect for sustainable development. It describes farming systems that are capable of retaining their productivity and value to society indefinitely and are environmentally sound, commercially competitive, resource conserving and socially supportive. Sustainable agriculture is defined as "a process in which the demands for its outputs-food, fibre and other services are met from farming practices that are economically efficient, environmentally friendly, and socially acceptable" (Legg, 1999). It can also be defined as "agriculture that is productive for the foreseeable future, competitive and profitable, conserve natural resources, protect the environment, and enhance public health, food quality and safety" (Puttaswamaiah S., 2005). Consultative Group on International Agriculture Research (CGIAR) defines, "sustainable agriculture is the successful management of resources to satisfy the changing human needs, while maintaining or enhancing the quality of environment and conserving natural resources". Sustainable agriculture involves multicultural, intercropping, use of farmyard manure, mulching and application of integrated pest management. Organic practices help in improving soil fertility and enhance the agricultural production in sustainable manner. Sustainable agriculture means raising food crops and related livestock species that is healthy for consumers and animals (such as dairy and other economically important animals). While it does not damage the environment, it provide fair wage to the farmers and support rural communities. The characteristics of sustainable agriculture include;

Conservation: Resources (land, water, soil and air) are vital to human. It should be available for future generations. The waste from sustainable farming should remain within the farm's ecosystem, and should not cause pollution.

Biodiversity: Farms raise different types of plants and animals, which rotate around the fields to enrich the soil and help prevent disease and pest outbreaks. Chemical pesticides should be use minimally and only when necessary, many sustainable farms do not use any form of chemicals (Gahukar, 2009).

Socially fair: Workers are treated fairly and paid competitive wages and benefits. They work in a safe environment and are offer proper living conditions and food. In 1977 and 1990, Farm bills introduced by USA government defines sustainable agriculture as an integrated system of plant and animal production practices having a site-specific application that will:

1. Satisfy human food and fibre needs,
2. Enhance environmental quality and the natural resource based upon which the agriculture economy depends,
3. Make the most efficient use of non-renewable resources and off-farm resources and integrate, where appropriate, natural biological cycles and controls,
4. Sustain the economic viability of farm operations, and
5. Enhance the quality of life for farmers and society as a whole.

The doubt about sustainable agriculture is that the definition is more like a philosophy or way of life than a legal binding or strict set of rules, and farmers can interpret the meaning differently (Gahukar, 2009).

Agriculture is essential to human survival and societal development. Population growth and economic development has increased the demand for agricultural products that has placed substantial pressure on agriculture and natural resources. This, in turn, has caused environmental pollution and ecological degradation. Agricultural sustainability has become a critical issue and so, in 1991 the concept of sustainable agriculture and rural development was introduce at Conference on Agriculture and the Environment in The Netherlands, held by the Food and Agriculture Organization (FAO) of the United Nations at Hertogenbosch (den Bosch). Its basic principles are to maintain a sufficiency of land for agriculture, to guarantee food security, to improve current living standards, to safeguard the development of future generations and to establish harmonious mechanisms for agriculture and economic development. Its general goals are to:

(i) Increase grain yield, ensure food security and eliminate famine,

(ii) Increase peasants' income, eliminate poverty and stimulate comprehensive agricultural development, and

(iii) Use and protect natural resources and the agricultural environment while improving the natural environment for present and future generations.

Literature Review

Hari Eswaran, 2007, in his paper show that sustainable agriculture is a management system for renewable natural resources that provides food, income, and livelihood for present and future generations while maintaining or improving the economic productivity and ecosystem services of these resources. In many developed nations, the concept of sustainable agriculture blends basic economic concerns, conservation, and maintenance or improvement of the resource base. The motivation is derived primarily from environmental and ecological concerns. Many developing countries still do not have detailed information on the resource base; consequently, database must be developed and techniques instituted to monitor resources (Eswaran, 2007).

Global population of approximately 6 billion in 2000 is expected to reach approximately 7.9 billion in 2025. Much of the increase in the population growth will also take place in the developing countries particularly in China and India in Asia, and in most countries of Africa. The question at this point is whether agricultural yield sustainability to meet the needs of approximately 7.5 billion people by 2030 without causing depletion of existing biodiversity and exhaustion of non-renewable resources is at all feasible, and if so, what the pathways are (P. C. Kesavan, 2007). In the context of the developing countries, particularly India and China, with a very large population, author recommends Integrated Farming Systems (IFS) as the framework for creating more food and livelihood (income). He has elaborated as and when properly designed and practiced, would ensure ecologically, economically and socially sustainable agricultural production and this could be an answer for serving million growing population.

Rajwinder Kaur, (2012) in her paper, aimed to investigate the free electricity as agricultural subsidy and its impact on Ground water table. Three districts of Punjab namely Ludhiana, Bathinda and Roopnagar were selected from each agro-climatic region. Study was based on primary and secondary data. Primary data was collected using detailed questionnaire. Study analyzed that cost of inputs reduced due to free electricity which result into enhanced income of farmers but negative impact on the environment. Electricity subsidy regressive as larger farmers are getting more benefit rather than small and medium farmers because of their larger land and have more than one connection and new pump sets etc. Author suggested flat rate of electricity rather than free electricity as electricity supply is irregular and farmers have to depend on diesel pump sets for irrigation, but same as the overexploitation of ground water (Rajwinder Kaur, 2012).

Reetinder Kaur, (2011) analyzed in her paper the interrelation between globalization and its effects on agriculture and as well on health in Punjab state. Study indicated that globalization leads to three types of alterations namely economic, socio-cultural and environmental. With the outcome of these changes, there is significant effect on health of residents of Punjab. There is noticeable rise in cancer, problems related to reproduction, mental disorders and kidney ailments. These three changes are inter-dependent and inter related. Green revolution is based on credit intensification and inputs such as synthetic fertilizers and pesticides (Reetinder Kaur, 2011).

Punjab State Action Plan on Climate Change, Government of Punjab (2014), report reveals that the Fertilizers consumption in Punjab is 235 kg per hectare which is 1.84 times higher than national average of 128 kg per hectare. Surface irrigation and ground water irrigation constitute major irrigation methods in Punjab. According to report, 72 % of the irrigation requirement is fulfilled through tube wells and rest from surface irrigation through canals and its distributaries. In terms of climate, trend of rise of temperature has observed in Punjab. Whereas, it experienced deficit in rain fall in terms of long term average. According to the last assessment of ground water carried out in 2009, the state faces deficit of 14.31 Billion Cubic Meter (BCM). State falls under the over-exploited category in terms of ground water extraction. High indebtedness found among the farmers in Punjab due to reduced farm yield coupled with declining productivity of soil and depletion of water table resulting in high cost of production. Abrupt changes in climate and increased variability's in rain fall is causing stagnation or even decline in wheat yield mainly in Ludhiana, Amritsar, Patiala, and Sangrur districts of Punjab (Government of Punjab, 2014).

Theoretical Underpinning

India comprises large number of workforce engaged in agriculture. As a result large wasteland area, forest area and grazing land are converted to cropland. Indian farmers have adopted the green revolution equipment like high-yielding variety seeds, fertilizers and pesticide, green revolution role in boosting the food grain output in India. While green revolution agriculture addressed mainly productivity issue, sustainable agriculture must address to multidimensional concerns like social, environmental and economic sustainability. Sustainable agriculture has become an interesting issue due to recent observations like, growth rate of output and productivity of crops have been falling down, increasing the problem of soil erosion, ground water table, increasing the uses of pesticides and fertilizers in several regions of the country, mainly in green revolution regions of India. Punjab state of India (known for pioneer green revolution state of India) has been experiencing a threat in the production of crops, problem of soil erosion, decreasing water table and high use of pesticides and fertilizers. Therefore, the analysis of agricultural practices of Punjab would address to the need for sustainable development.

Selected Study Area

The state of Punjab comprises an area of 50,362 square kilometres which is 1.5 per cent of the total geographical area of the country. It extends from the 29.30° to 32.32° north and longitudes 73.55° to 76.50° east. The subtropical location leads to variation in temperature. Punjab experience three main seasons: hot season, rainy season and cold season. The monsoon season provides most of the rainfall in Punjab. The climate of Punjab is influenced by the Himalayas in the north and the Thar Desert in the south and south east. The state receives only 61.9 cm (Average) rainfall, of which 75 per cent is received during monsoon months. According to the census 2011, the state of Punjab has 2, 77, 43, 338 persons. This is 2.30% of the total population of India and Punjab has 16th rank amongst all 29 states of India. The decadal change increase in population from 2001 to 2011 is 13.89 per cent.

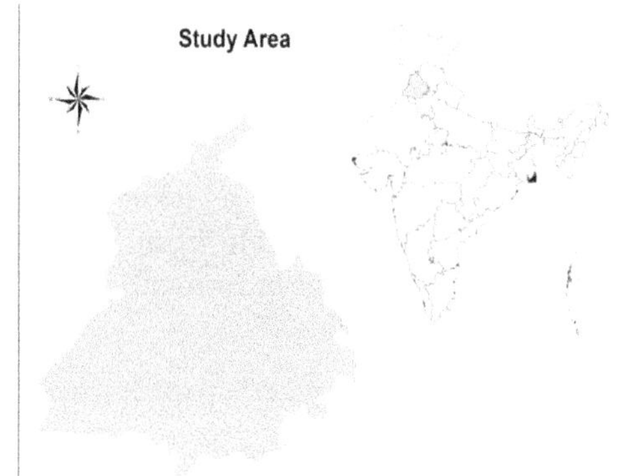

Objectives of Study

The present paper, a review and highlight the status of sustainable agriculture development scenario in India with special reference to Punjab; review the Indian national agriculture policy for sustainable agriculture and sustainable development goal of Punjab for Agriculture (SDG-2030). The present status of natural resources management practices for sustainable agriculture development and discuss the drawback of conventional agriculture in term of instability in production and diminishing natural resources. In this major focus to, review the importance and needs of sustainable agriculture in Punjab and also review the guideline of Punjab Government, to achieve Sustainable Agriculture Goals also known as Vision 2030. Suggestion to promote the sustainable agriculture practices for long time sustainability for agriculture.

Research Methodology

The present study is largely based on secondary sources of data that is collected from journal articles, government reports, periodicals, newspapers, authentic web sources and agriculture policy related data collected from Punjab Agriculture Department, Chandigarh. Water Level related data collected from Directorate of Water Resources, Chandigarh. Over-all past performance of the state since one decade has been review and assessed in agriculture sectors. All the showing maps are prepared with the help of ArcGIS Geospatial Technique, and using ISRO's Geoportal Bhuvan.

Results and Discussions

Agriculture, with roughly half of India's work force, is still engaged in agriculture for livelihood. India has a higher priority to agriculture and in achieving the goals of sustainable agriculture is important for sustainability in agriculture production and environment. Indian National Agriculture Policy for Sustainable Agriculture aims to attain; 1. Growth rate is excess of 4% per annum in the agriculture sector, 2. Growth that is based efficient use of resources and conserves soil and water, 3. Growth that is widespread across regions and farmers, 4. Growth that is demand driven that caters to domestic market and maximize benefits from exports of agricultural products in the face of the challenges arising from economic liberalization and globalization and 5. Growth that is sustainable technologically, environmentally and economically.

Land Use Pattern of Punjab

The land use classification for the year 2007-08 to 2016-17 (table 1) shows that the net sown area was 41.87 lakh hectares in 2007-08 and declined to 41.30 lakh hectares in the year 2016-17. This could relate to urbanization, resulting into decreasing in gross cropped area in the state from 78.70 lakh hectares to 78.29 lakh hectares. The area under permanent barren and uncultivable land has been increase at, this more than double in 2016-17.

Table 1: Land Use Pattern in Punjab (000, hectares)

Area /Period	2007-08	2013-14	2016-17
Geographical area	5036	5036	5033
Forests	287	258	256
Barren and un-cultivatable land	24	51	52
Net sown area	4187	4150	4130
Net area sown as percentage to total area	83	82	82
Area sown more than once	3683	3720	3699
Gross Cropped area	7870	7870	7829

Punjab soil and water are important natural resources for agriculture where the presence of alluvial soil (Krishna H. Shukla, 2015) is boost for Punjab agricultural development. The soil profile of the state as followed; the Nitrogen and phosphorus are also found low to medium in most parts of the state soils. The soils in general are medium to high in available potassium. The mass production of paddy and wheat has exhausted the vital nutrients of the land of the state. The soils contain sufficient calcium and magnesium. It is being felt seriously that there should be crop diversification in the state otherwise the land is bound to get unproductive because of the paddy and wheat cycle which is repeating into the very vitals of the state land. Thus, it is more important to preserve existing cultivated areas in the state from degradation due to water logging, soil salinity and solidity, besides soil erosion due to intensive cropping and its attended manifestations. Figure No. 2 describes the status of soil erosion and salt affected soil in Punjab. The problem of water logging is particularly acute in south-western districts (Ferozepur, Fazilika,

Muktsar and Faridkot) of the state occupying lower topographic positions. Introduction of salt resistant crops and good drainage system may overcome this problem to some extent.

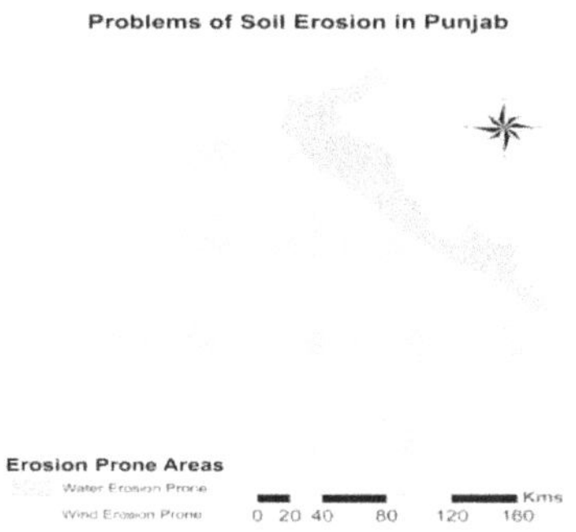

Ground Water Level of Punjab

Change in cropping Pattern after the inception of green revolution in Punjab farmers shifted from low water consuming crops to high water requiring crops such as paddy crop irrespective of soil conditions and rainfall quantity. Large part of the Punjab state for irrigation needs, depending on ground water; recently the table of ground water is going down by 30 cm per year in Punjab. It is declining in 7 % of the areas of the state where ground water quality is good and canal water supply is limited. A critical water table depth of below 60 feet has reached in 60 percent of the state. Fig no. 3 shows the temporal change in water level in 2007 and 2017. There has been a phenomenal rise in area under paddy from 3000 km^2 in 1975-76 to 12000 km^2 in 2011-12. The irrigation water requirement of this increased paddy crop are fulfilled by ground water and second one is increase the Number of Tube wells has increased from 0.20 million in 1975-76 to 0.73 million in 2011-2012. This has resulted into excessive over exploitation of ground water (RAO, 2002).

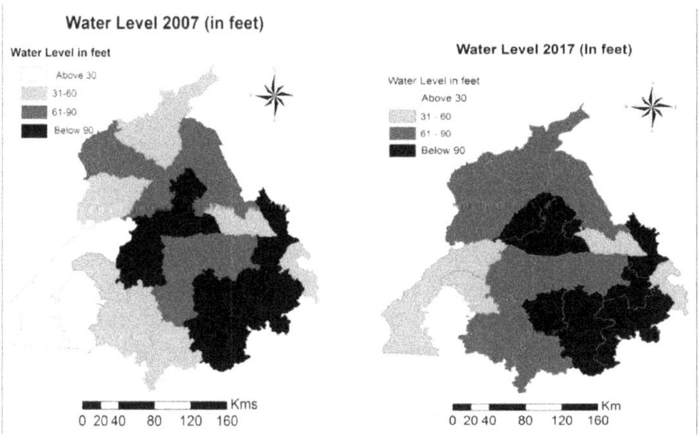

Fig No. 3

State Crop Intensity profile: Cropping intensity signifies the extent of multiple cropping. The cropping intensity has been growing continuously in the state of

Punjab and it has become a common feature now to get two crops from the same field. The strategy is to maintain the crop intensity for natural resource management like soil, ground water, for achieving to sustainable development goal 2030. For sustain cropping intensity government ensure that agriculture universities and research centres develop new techniques and technologies and timely supply to farmers on low costs. Table 2 highlights the level of cropping intensity from the year 2007-08 to 2016-17. It signifies that the cropping intensity has increased marginally in state from 188 percent in 2007-08 to 189 percent in 2016-17. However, this high intensity of cropping is a clear indicator that that in Punjab state the vertical expansion of area in future has become increasingly limited.

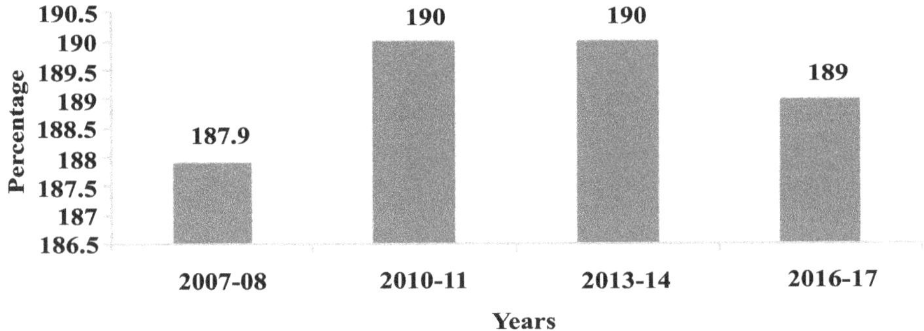

Fig, 4: Cropping Intensity in Punjab

Source: *Statistical Abstract, Punjab*

Irrigation Status of Punjab

Water is the elixir of life and agriculture too. The state has an effective network of irrigation facilities and about 98 percent of the net sown area in the state is irrigated. The following table no. 2 highlights the irrigation facilities of the state:

Table No-2: Gross Cropped and Irrigated Area in Punjab (000, hac)

Year	Gross cropped area	Gross Irrigated Area	Percentage
2007-08	7870	7689	97.7
2010-11	7882	7723	98.0
2013-14	7870	7771	98.7
2016-17	7829	7757	99.1

Source: *Statistical Abstract, Punjab*

Over the time, canal irrigation has been declining whereas tube well irrigation has been increased and about 73 percent of the total irrigated area is being irrigated by underground water pumped out by about 13.80 lakh tube wells. This is mainly due to the availability of cheap credit and free supply of electricity in the state. However, this has led to the decline the level of the underground water which is at present a serious concern for the state.

Status of Pesticides use in Punjab: The use of pesticides and insecticides played an important role in agricultural production; use of chemical pesticides during green revolution era has been primitive in nature. The policy supports in favours of HYVs naturally, the application of chemical pesticides. Later on, these chemicals were used in a highly unscientific way, mostly in Punjab and Haryana state, starting from the choice of chemical to the application practices, timing and even necessity. (Indira Devi, 2017) To reduce the pesticide use, it is necessary to identify the high pesticide using areas by analyzing farm gate samples of vegetables, basmati rice, fruits and other crops for pesticide residues. To minimize the build-up of pesticide residues and to keep the same within scientifically acceptable limits appropriate strategies should be developed and monitored. The farmers should be accordingly educated to follow such recommendations for the produce to be within the pesticide safety limits. A state-of-the-art laboratory should be established in the State for residue analysis of pesticides and heavy metals in the crop produce (State, 2013). Cropping intensity and fertilizer use also increased the problem of weeds. The consumption of pesticides and insecticides is shown in the Figure No. 5 as under, the use of pesticides in Punjab, as per data table, 5900 metric tons pesticide use in the year of 2007-08 as follow in 2016-17, it is near about 5843 metric tons.

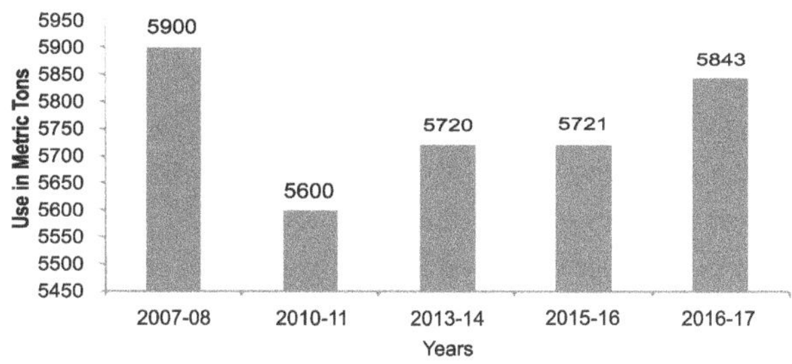

Fig. 5: Use of Pesticides in Punjab (In Metric Tons)

Source: *Agriculture at a Glance, Department of Agriculture, Government of Punjab, Chandigarh.*

Agriculture Development in Punjab

Punjab has environmental, institutional and technological factors that reflect forward agricultural development evolving since independence (Chadha, 1986). Despite having less than 2 % of the country's area, the state of Punjab is surplus state in terms of food grain. It contribute significant portion (34 percent for rice and 75 percent for wheat) of grains to the central pool for Government distribution (MCA, 2017-18). Punjab agriculture is dominated by a rice-wheat production system accounting for almost 80 % of the cropped area and over 85 % of the gross value of crop output. The specialization in these two food grains can be attributing to the large-scale adoption of high-yielding technologies, favourable policies, and massive investment in the irrigation and power sectors. The predominance of this cropping system was responsible for alleviating poverty and bringing prosperity to the state during the 1970s and 1980s (PAU, 2014).

The balance between rainfall and evaporation is important as the agro climatic condition divide the area based on climatic parameters relevant to agriculture, indicating areas that are climatically suitable for different crops. Punjab can be divided into three regions: Zone I (the sub-mountainous region), Zone II (the central belt) and Zone III (the south-western cotton belt).These agro-climatic zones are unique in terms of climatic conditions and suitability for the crop. However, crops are not cultivated according to natural climatically conditions. For example zone, south-western zone which is not suitable for rice cultivation (according to agro-climatic conditions), south-western zone is suitable for cotton-wheat crops. However over the last decade there has been fall in area under the cotton crop and at the same time area increase under the rice. This change has mainly effected on the use of underground water and soil properties that could pose severity for long time sustainability of soil properties, crop production and for the natural recourses. If this conventional practices continue, so we will never be able to achieve the goal of sustainable agriculture (Singh, 2004). For development of sustainable agriculture practices government of Punjab introduces Sustainable Development Goal, for agriculture (SDG-2030) also known as Vision 2030: Goal of Punjab Government: "Zero Hunger: End Hunger, Achieve Food Security and Improved Nutrition and Promote Sustainable Agriculture" this goal is part of United Nation framework of Sustainable Agriculture and Indian National policy of Sustainable Agriculture.

The Major Challenges and Strategies of the Policy (Vision 2030)

According to (Department of Planning, 2016) the contribution of agriculture in total Net state Domestic Product (NSDP) is declining very fast within last five years it declined by 12 percent points and the NSDP originating in agriculture grew by just 0.33 percent per annum during this period. The average yield of crops, in Punjab shows in Figure 3. At present rice yield is growing at the rate of 1.15 percent per year and wheat yield at the rate of 0.75 percent per year. The rice and wheat yield in Punjab being already the highest in the country, it will require considerable effort to maintain the existing growth rate of rice and wheat yield with sustainable development growth practices.

Sustain Current Growth Rate of Rice and Wheat Yield

Agriculture production in Punjab is dominated by rice and wheat crops, these crops are cultivated on about 80 percent of cropped area and contribute 85 percent of the total value of crop production. Cotton is the only other important crop that is cultivated on about 4 percent of cropped area and contributes about 5 percent of the value of total crop production. The remaining 15 percent of value of crop production is contributed by maize, sugarcane, pulses, oil seeds etc. (Department of Planning, 2016).

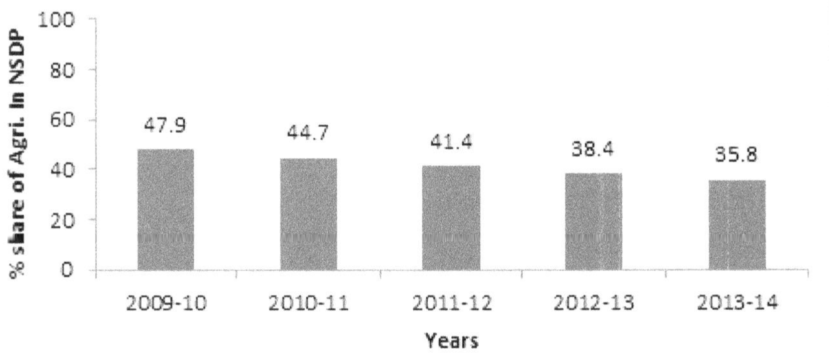

Fig. 6: Contribution of Agriculture in total NSDP (2004-05 prices) in Punjab.

Source: *Handbook of statistics on Indian Economy, 2014-15, RBI.*

Strategy to Maintain: For achieving to increase low yield areas, the low yield villages and low yield farmers need to be identified and the causes of their low yield will be found and policy intervention will be implemented to remove these causes and help the low yield areas, villages, and farmers to raise their yield to the Punjab average level. The second component of the strategy is to ensure that all farmers use only university/experimental farm produced seed of rice and wheat. The agricultural department will ensure that adequate supply of such certified seeds is available to farmers at the sowing time. The third component of the strategy is to ensure the timely supply of adequate power, fertilizer, credit and other inputs, so that rice and wheat are sown during the period that results in highest yield. Other things remaining the same, the timely sowing results in 10 to 15 percent increase in yield (Department of Planning, 2016).

Develop Soil and Water Management practices

There is an over exploitation of water resources in several regions as indicated by rapid decline in ground water levels. There is depletion of ground water in about 110 blocks out of total 146 blocks, Likewise, intensive cropping using varieties having higher productivity have caused mining of soil nutrient reserves, leading to depletion of soil fertility and emergence of multi-nutrient deficiencies of macro and micro nutrients in the state. In the south western districts of Punjab, water logging and saline water are posing problems in crop cultivation (ENVIS, 2014).

Strategies to Maintain: In order to Punjab context the address challenges, need for concerted efforts to improve water and nutrient use efficiency and enhance soil health. Its, imperative to reduce ground water use by developing and refining water saving technologies in crop production such as short duration rice varieties, direct seeded rice, drip irrigation and fustigation systems, laser levelling, mulching, in-situ farm rain water conservation through construction of water harvesting structures and ground water replenishment through recharge structures have to be promoted. In some countries, like USA, Argentina, Australia and many-more, used sensor-based technologies are proving useful in the efficient management of natural resources. Precision agriculture like laser land levelling and soil mapping through GPS are the outcome of such approaches. In south-western Punjab, use technologies for conjunctive use of brackish and good quality waters as well as management of brackish water will be further refined and rice genotypes having tolerance to salinity will be identified. Bio-drainage technology using fast growing tree species such as *Eucalyptus* and salt tolerant crop varieties such as that of rice will be developed for water-logged areas. Excessive use of nitrogen fertilizers is a serious issue. There is urgent need to adopt appropriate soil, crop and land management practices that help to mitigate ill effects of climate change on soil environment, reduction in soil degradation, maintaining soil quality (Department of Planning, 2016). More research is required to design appropriate water and soil management strategies in order to mitigate ill effects of climate change on soil environment and obtaining better crop yields (Department of Planning, 2016).

Findings and Suggestions

The concepts of sustainable agriculture reflect an introductory phase in Punjab. Finding reveal that the current agriculture scenario deviating from sustainable development goal. Current agriculture practices does not maintain the natural resources which include over-exploitation of natural resource like soil, and water by using excessive chemical fertilizers and pesticides. The practices are not scientifically recommended for maintaining ecological and soil fertility balance. The resultant factor could be observed in increased land-use, degradation of the soils, and over-exploitation of ground water resources, surface water and related ecosystems. Growing population coupled with rapid urbanization, demand for agriculture and food products and thus put severe pressures on land and water resources to meet this demand, rapid urbanization pressure (shows in table no. 1) on land demand effect of this area under agriculture is decrease day by day.

In order to meet food demand of the population and maintaining the ecological balance and environment, drastic changes in agricultural practices are required. Therefore, products, technologies and services must be developed to enhance production capacity while protecting and restoring natural resources. Vision 2030 of Punjab set three goals; to sustain Current growth rate of rice and wheat yield, but on the basis of data as shown (Fig. no. 6) the current status of the yield production is not stable, it will require considerable effort to maintain the existing growth rate of rice and wheat yield with sustainable development growth practices for achieving the timely target, secondly government develop new soil and water management practices to sustain the natural resources for long time availability, of the natural resource for agriculture but the present status of soil and water resources is very critical, ground water table is below 90 feet as (show in Figure No. 3) in many districts of Punjab. The problem of Soil erosion is also very serious in the Sothern-western parts of state as (show in Figure No. 2). The analysis has pointed out to the need for a detailed look on the pesticide-use pattern, distribution systems and regulatory mechanism at a micro level. The attainment of green growth strategies and sustainable development goals necessitates the adoption of safe production practices in agriculture that could be ensure through scientific management of supply chain systems.

References

Chadha, G. K. (1986). *The State and Rural Economic Transformation: The Case of Punjab 1950-1985*, Sage Publications, New Delhi, p.370

Chasan, R. (1992). The National Research Initiative: Enough Support for Sustainable Agriculture, *The Plant Cell*, 4(1), 3-4.

Eswaran, H. (2007). Sustainable Agriculture in Developing Countries. In *Agriculture and the Environment* (pp. 199-203). Washington, DC.

Europrean commission (2013). *Sustainable agriculture for the future we want*. https://ec.europa.eu/agriculture/sites/agriculture/files/events/2012/rio-side-event/brochure_en.pdf

FAO (2017). *The State of Food and agriculture.* United Nations

Gahukar, R. T. (2009). Sustainable Agriculture in India: Current situation and future needs. *International Journal of Agricultural Sciences*, 5(1), 1-7

Kesavan, P. C. and Swaminathan, M. S. (2008). Strategies and models for agricultural sustainability in developing Asian. *Philosophical Transactions of The Royal Society*, 363, 877-891

Shukla, K. H. and Dwivedi, N. U. (2015). Sustainable development in agricultural sector in India. *The Business & Management Review*, 5(4), 220-222

Shukla, O. P. (2017). Relevance of Sustainable Agriculture in India. *Journal of Management Science, Operations & Strategies*, 1(3), 28-32

Singh, J., & Nandal, D. (1993). Sustainability of agriculture on degraded irrigated lands in Haryana. *Indian Journal on Agriculture Economies*, 48-56.

Smit, B. and Smithers, J. (2016). Sustainable Agriculture: Interpretations, Analyses and Prospects. *Canadian Journal of Regional Science*, 499-524.

Takle, S. R. (2016). Sustainable Agriculture development in India, *Agricultural Science Research Journal* 6(11), 281-286

UNDP (1987). *World Commission on Environment and Development: Our Common Future* United Nation

GoI (2016). *Annual Report* (Government of India).

GoI (2017-18). *Annual Report* (Government of India).

GoP (2008, 2014, 2016). *Statistical Abstract.* Chandigarh Economic and Statistical organisation

GoP (2014). *Agricultural Development in Punjab*, Ludhiana (Government of Punjab).

GoP (2016). *Punjab Vision Document 2030*, Chandigarh Institution for development and Communication

Punjab State Council for Science & Technology, (2014). *State Environment Report*, Chandigarh

Gujarat Institute of Development Research, (2005). *Promoting Sustainable Agriculture: In India and Canada*, Gota, Ahmadabad

10

Role of Education in Sustainable Agricultural Marketing Practices in Indian Horticulture

Dipika Basu, Arun Kumar Nandi, Sulata Hembrem, Pratha Pratim Roy and Uttam Haldar

Introduction

Increasing demand for horticultural products is an opportunity to revitalise agricultural growth and augment income of the farmers through diversification of agriculture towards horticultural crops in India (Birthal et al, 2008). Horticulture-led growth has considerable potential to create employment opportunities for small and marginal farmers to enhance their income from farming. There is also a huge scope of employment generation in the non-farm sectors like transport, retail and wholesale trading, storage and processing industries, packaging and exports etc., particularly for the vast number of unorganised and low skill labourers in India. Due to strong backward and forward linkages of horticulture sector to non-farm sectors, rational use of agricultural resources and sustainable management of both inputs and outputs of horticulture is very crucial for transforming agricultural growth to sustainable economic development.

A very complex and diversified production-distribution-consumption links of agri-products challenges to sustainable management of the system of agricultural production and marketing. Education is one of the most important instruments for sustainable use of natural and human resources in agriculture. It empowers farmers to change the way they think, innovate and work for sustainable agriculture. Education helps people to develop knowledge, skills, values and behaviour which are indispensable for sustainable development. UNESCO aims to improve access to quality education for sustainable development at all levels and in all social contexts. Education and agricultural extension services can play a significant role in the management of human resources, technology and natural resources involved in the system of agriculture towards sustainable future and environment. In this context, the present study is an attempt to assess contribution of education to employment and earnings from agriculture with particular focus on Indian horticulture production and marketing systems.

Literature Review

The major changes that are taking place in Indian agriculture are: (i) shrinking resource base due to inappropriate and unscientific use of land and water, (ii) changes in demand and consumption pattern in favour of fruits, vegetables, meat, fish, eggs and dairy products, (iii) there is a decreasing trend in per capita cereal consumption, (iv) there is a shift of farming systems from food grains to non-food grains cultivation which needed different types of support systems like training, extension services, problem-solving consultancy, marketing advice etc., (v) public investments in agriculture is declining in real terms, and (vi) country's policy of economic openness results new opportunities as well as new threats to agriculture. Particularly, Indian agriculture faces a big challenge under economic globalisation and liberalization (Sulaiman & Van den Ban, 2000).

Education is one of the most important instruments for sustainable management of agricultural resources for sustainable agricultural growth and development. It is increasingly realized that knowledge is an important input for efficient farming. In the present era of globalisation, agriculture is becoming an increasingly recognised as knowledge and information-driven enterprise. Education in general and agricultural research and extension education in particular are not only important to increase agricultural production, productivity and profitability but also very crucial for improvement of social and economic life of vast rural people associated with the farming system. Agricultural extension education and advisory services are a vital element of the array of market and non-market entities and agents that provide critical flows of information that can improve farmers' and other rural peoples' welfare (Anderson, 2007). But it is very unfortunate that the existing knowledge and information about how to use optimal combination of inputs, know-how, land management methods, and how to process, and market agricultural commodities remain inaccessible to a vast number of marginal and small farmers in India (Babu & Joshi, 2014). This is a matter of grave concern to the policy makers.

NSSO survey results (2003) regarding situations of agricultural households showed that 60 per cent of the farmer-households did not access any information on modern technologies in India. In a study it is found that: (i) one-third of the farmers obtained extension services from progressive farmers and input dealers, (ii) about 29.3 per cent farmers used radio, television and newspapers, (iii) the contribution of public sector extension system is about 10 per cent, and (iv) only 0.6 per cent of the farmers was accessed private and NGO extension services. The proximity (33.7 %), assured quality (21.1 %), sole option (20.6 %), and timely availability (13.7 %) were the main reasons for the choice of such information (Babu *et al.*, 2012). In another study it is also found that the service delivery by public-sector extension workers was lowest for small farmers (4.8 %) compared to large farmers (12.4 %) (Adhiguru *et al.*, 2009). Particularly, with the development of new technologies our agricultural extension system faces important challenges in the areas of relevance, accessibility, accountability, efficiency and sustainability.

Several studies have been made to establish the relation between the income and education both in India as well as outside India. As far as India is concerned, there are some studies which are based on NSSO data (Duraisamy 2002; Dutta 2006; Kingdon and Theopold 2008; Madheswaran and Attewell 2007). Studies suggest returns to education are higher for lower levels of education (e.g., primary) and decline with the level of education. This is due to the low cost of primary education relative to other levels of education and considerable productivity differentials between primary graduates and illiterate persons. But in this study an attempt has been made to show the impact of education among the rural people and to substantiate the role of education as well as agricultural training for the improvement of the productivity and earnings among the people engaged in agricultural activities in West Bengal.

The role of education is considered to be important for earning higher income level. But the impact of education is not the same among the agricultural and the non-agricultural sector. So when it comes to education the important thing which is important to understand is that whether it is increasing the earnings for all the sectors or it is paying attention to some specific sector. If education does not increase the earnings of the agricultural workers then investment in education is not viable for the people who are engaged in the agricultural sector. On the other hand, if it is increasing the level of earnings then what level of education is contributing the most is important question. Moreover, not only education but useful and effective training in agricultural practice as well as the management of the sustainability of the same is also very important.

Objectives and Data

The objectives of the study are: (i) to examine role of education and agricultural extension services in sustainable agricultural practices in West Bengal with special reference to Indian horticulture; (ii) to estimate returns to education on earnings of the persons belong to agricultural households in West Bengal; and (iii) to suggest some policies for sustainable management of agricultural resources and sustainable agricultural growth for sustainable future.

The study is based on secondary data collected from different government sources like the National Horticultural Board, National Sample Survey Organisation, Government of India and Bureau of Applied Economics and Statistics, Govt. of West Bengal. To assess the impact of education on income of rural agricultural households in West Bengal we have considered unit level data extracted from the 70th round of NSS on Situation Assessment Survey (SAS) of Agricultural Households, 2013.

We have considered the people engaged in agricultural activities residing in the districts of Darjeeling, Jalpaiguri, Koochbihar, Uttar Dinajpur, Dakshin

Dinajpur, Maldaha, Murshidabad, Birbhum, Barddhaman, Nadia, North 24 Pargana, Hugli, Purba Midnapore, Paschim Midnapore, Bankura, Puruliya and Howrah. The target group of the study is basically the rural people engaged in agricultural activities of West Bengal.

Estimation Methods

For estimation of the returns to education we have taken two groups where the first group consists of the people who worked in household enterprise (self-employed: own account worker, employer, worked as helper in household enterprise (unpaid family worker), worked as regular salaried/ wage employee, worked as casual wage labour: in public works other than MGNREGA works, in MGNREGA works and the people who worked in other types of work. Since the study is based on the rural people of the districts of West Bengal as a result the study is basically of the agrarian people. The sample size for the first group includes 1609 persons. The second group of the sample includes the people who are regular wage or salaried people only. This is a homogeneous group consisting of people with agricultural background. The sample size is small with a number of observations equal to 477.

For the estimation of rate of returns to education on earnings from the agriculture we have used the earnings function method (Psacharopoulos 1981, 1994). As the elaborate method requires entire information regarding age-earnings as well as cost of education profiles by educational level, the elaborate method (Becker, 1964) is not used. This study estimates returns to education based on the earnings function method, which is also known as human capital earnings function or 'Mincerian' method of the people engaged in agricultural activities. An interesting aspect of Mincer's model is that the time spent during schooling is a key determinant of the earnings. The basic 'Mincerian' earnings function (Mincer, 1974) is given as: $\ln Y_{ij} = \beta_0 + \beta S_{ji} + \mu_1 E_i + \mu_2 E^2_i + u_i$

The dependent variable used in the analysis of the study is log of income and the independent variable incorporating the human capital variable can be extended by levels of education of the rural people where dummy variables are used for different levels of education using standard Mincerian earnings function named after Jacob Mincer.

Model Specification: $\ln Y_{ij} = \beta_0 + \sum \beta_{ji} S_{ji} + \mu_1 E_i + \mu_2 E^2_i + u_{i,}$

S_{ji}= i[th] individual completing j[th] level of education, E= Potential Experience in the labour market, E^2_i =Squared term of potential experience of the labour market where potential experience is defined by the difference between age and years of schooling less 6 i.e. the age at which a person gets enrolled in school.

This model is purely a model of heterogeneous model of returns to education (Stearns, 2000) since the level of earnings is different for different level of education and education is not treated as continuous variable rather discrete. So, with the growth of level of education the benefit of education also goes up.

However, we also want to check the impact of some demographic factors as well as social characteristics and the impact of migratory trend for those who have spent at least 15 days continuously outside their native land, locational effect reflected by region (dummy variable) and finally impact of agricultural training on the level of income. So, the equation becomes

$$\ln Y_{ij} = \beta_0 + \sum \beta_{ji} S_{ji} + \mu_1 E_i + \mu_2 E^2_i + \mu_3 G_i + \mu_4 T_i + \mu_5 M_i + \mu_6 R_i + u_{i,}$$

G_i= Gender male or female, T_i= Trained or Untrained, M_i= Migrating outside the native land

R_i= Region whether Gangetic West Bengal or Non-Gangetic West Bengal (including terai and Jungle Mahal economy).

The equation shows that the log of income depends on the educational level of individuals, experience, squared experience, Gender, Training obtained, Staying pattern of the individual and finally the Region.

In this context, it is important to understand the geographical diversity of West Bengal since we are concentrating on the rural people and mostly on the households who are engaged in agricultural activities. So, the topography of the area under study is important. As a result a dummy variable is incorporated which is denoted by R_i and it takes the value 1 for the districts which belong to gangetic West Bengal and takes the value 0 for the districts which are under terai, dooars and jungle mahal more precisely non-gangetic part of West Bengal. The gangetic part of West Bengal is fertile and the productivity of agricultural crops is more compared with the non-gangetic part of the state. So the coefficient of R_i represents the differential effect due to location.

Similarly, the co-efficient of gender represents the differential effect of being male or female, the co-efficient of training represents the differential effect of being trained and finally the co-efficient of the variable M_i represents the differential effect of staying outside the native land for more than 15 days.

In the next segment a comparative analysis is done regarding the mean earnings of the people of the representative districts. Also, there is availability of data regarding the individuals training in agricultural activities and motivation towards migration. An attempt has also been made to capture the relationship between training in agricultural activities and motivation towards migration and their impact on income. The Table-1 below describes the nature of the variables used for regression.

Table 1: Description of Variables

Variable	Description	Base Category
	Dependent Variable	
Log wage	Natural log of wage in Rs. to estimate the earnings function.	None
	Explanatory (Independent)variables	
Human Capital Variables		
Educational Level	An individual belongs to one of the following educational level: illiterate and literate below Primary, Primary, Secondary, Higher Secondary and Graduate & Post Graduate. It is assumed that an individual spends 0, 4, 6, 2, 3 and 2 additional years respectively in these educational levels. (5 Dummies: Primary, Secondary, Higher Secondary, Graduate & Post Graduate)	Illiterate and literate below primary (S_{ji}=0)

Experience (E)	Potential experience (Proxy for actual labour market experience) in years is defined as age minus years of schooling minus 6	None
Experience^2 (E^2)	Square of experience	None
Demographic Variables		
Gender	Sex of Individual (Male or Female) Dummy Variable	Female (G_i =0)
Regional	Place of Residence: Gangetic West Bengal Versus Non-Gangetic West Bengal (Terai, Dooars and Jungle Mahal Economy), Dummy Variable.	Non-Gangetic West Bengal (=0)
Training	Agricultural training taken by individuals versus untrained, Dummy Variable	Untrained (T_i =0)
Stayed outside	Stayed outside for at least 15 days, Dummy Variable	Stayed inside (M_i =0)

Semi logarithmic regression equation is used to estimate annual average compound growth rate of the relevant variables in the study.

Results & Discussions

India's Position in World Horticulture

Horticulture is a subset of agriculture sector which covers wide varieties of crops from vegetables, fruits, and flowers to spices, honey and different plantation, aromatics and medicinal crops and plants. It is a rapidly growing sector in the country. The National Horticulture Mission scheme has been launched in 2005-06 for the holistic growth and development of horticulture sector. The area under horticulture crops in India has increased from 18.5 million hectare in 2005-06 to 24.9 million hectare in 2016-17 with a growth rate of 2.9 per cent per year and the corresponding horticulture production has increased at the rate of 6.7 per cent per year; from 166.9 million MT to 300.6 million MT during the same period. The growth rates of exports of fresh vegetables, fresh fruits and floriculture in value terms are significantly high of 9.03 %, 15.31 % and 5.64 % respectively as compare to the overall growth of India's agri-exports (2.37%) during 2011-12 to 2017-18.

Table 5.1 India's Position in World production of Fruits & Vegetables, 2014

Item	Production (million tonnes)		India's		Country ranked 1st.
	India	World	% Share	Rank	
Vegetables & Melons	127.0	1169.0	10.8	2nd	China
Okra	6.3	9.5	66.3	1st	India
Potatoes	46.0	382.0	12.0	2nd	China
Tomato	19.0	171.0	11.1	2nd	China
Onion (dry)	19.0	88.0	21.5	2nd	China

Item	Production (million tonnes)		India's		Country ranked 1st.
	India	World	% Share	Rank	
Cabbages & other Brassicas	9.0	72.0	12.5	2nd	China
Cauliflower & Broccoli	8.5	24.0	35.2	2nd	China
Brinjal	13.5	50.0	27.0	2nd	China
Fruits excluding Melons	88.0	689.0	13.0	2nd	China
Banana	28.0	114.0	24.5	1st	India
Mango, Mangosteen &Guava	18.0	45.0	40.0	1st	India
Lemon & Lime	2.8	16.2	17.2	1st	India
Papaya	5.6	12.6	44.4	1st	India
Apple	2.5	84.6	3.0	4th	China
Pineapple	1.7	25.4	6.8	7th	Costa Rica
Oranges	7.3	72.3	10.1	3rd	Brazil
Grapes	2.6	74.5	3.5	8th	China
Major citrus (total) fruits	11.2	139.8	8.0	3rd	China

Source: *Horticultural Statistics at a Glance, 2017, DAC & FW, GOI.*

An analysis of data in Table 1 shows that the position of India in world horticultural production is the second largest after China. It is proud of us that India has secured 1st position in the production of Okra, Banana, Mango, Mangosteen, Guava and Papaya among the countries in 2014. Therefore, it is a bright side from the production point of view. But, if we examine data on land productivity of vegetables and fruits, it can be said that the Indian horticulture productivity is quite low in comparison to other countries (Figure 1). India's vegetables productivity is found to be 14.87 tonnes per hectare while the said productivity in USA is 33.30 tonnes per hectare. In China it is also quite high of 24.04 tonnes per hectare as compared to India. Thus, there is great potentiality to improve productivity of Indian horticulture. So, the prospect of Indian agriculture lies in the improvement in yield rates of different crops. Education, training, agricultural extension services etc. are the crucial inputs for growth and development of modern agriculture in India.

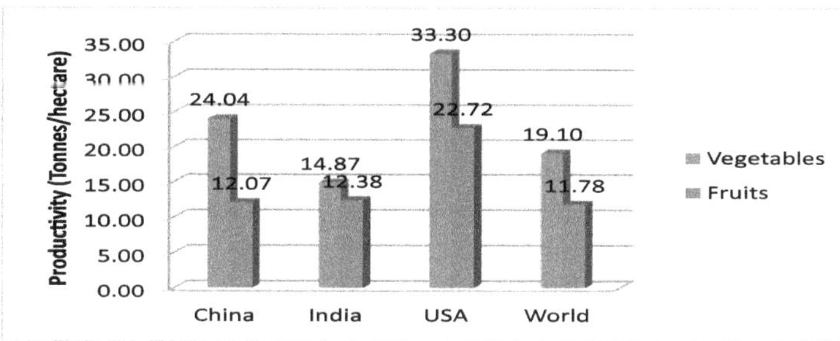

Fig. 1: Productivity of Vegetables & Fruits in China, India, USA and World, 2014

Table 2 shows the growth rates of traditional principal crops and horticulture crops (vegetables and fruits). It reveals that the growth rates of traditional crops are estimated to be less compared with the horticultural crops during 1961 to 2016. Secondly, among the horticultural crops, if we compare the pre and the post-reform periods, the rate of growth of horticultural crops is observed to be greater in the post-reform period than in pre-reform period which implies prospects of horticulture sector in India.

Table 5.2 Growth rates of Area and Production of different crops in India, 1961-2016.

	AACG	AACG	AACG
Growth of Area & production	**1961-1990**	**1991-2016**	**1961-2016**
Crops	**Pre-reform**	**Post-reform**	**Total**
Paddy area	0.63	0.06	0.41
Paddy production	2.67	1.43	2.31
Wheat area	2.50	0.93	1.47
Wheat production	6.22	2.03	4.06
Fruits area	1.85	4.01	2.91
Fruits production	2.55	4.57	3.65
Vegetables area	2.05	2.61	1.79
Vegetables production	3.47	3.88	3.43

Note: *a=Area in 000 hectare and p=Production in 000 tonnes. AACG = annual average compound growth rate (%).*

Source: *FAO Data base.*

Therefore, the major concern is to find out the factors which are responsible for the development of the people engaged in the agricultural economy in India. Diversification of agriculture in favour of horticulture has great potentiality to proper use and management of agricultural resources. The management of human and physical sources and effective optimisation techniques and policies might increase the productivity. In the present context of knowledge-driven agriculture, the impact of education is therefore needed to be carefully analysed.

West Bengal in Indian Horticulture

Vegetables occupy a prime place in West Bengal horticulture. The area under vegetables is about 76 per cent of total area of horticulture in West Bengal followed by fruits area (13 per cent), and flowers (2 per cent) in triennium crop year ending of 2016-17. Others horticulture plantations and crops share about 9 per cent of total horticulture area (Figure 1). In India, the shares of vegetables, fruits, and flowers in total area under horticulture are 41 per cent, 26 percent and 1 percent respectively in 2016-17 (Figure 2).

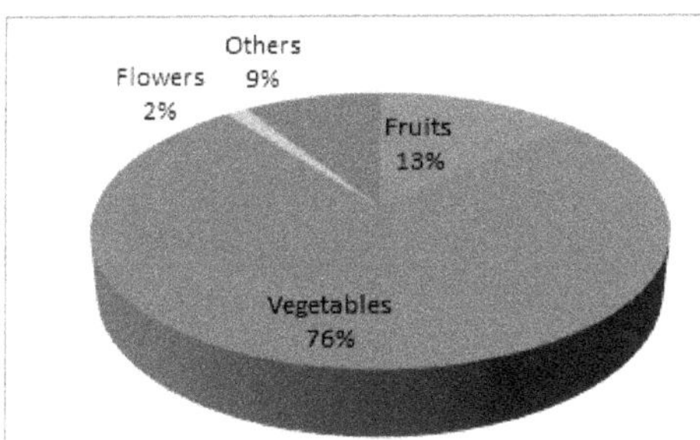

Fig. 2: Distribution of Area under horticulture in West Bengal (TE: 2016-17)

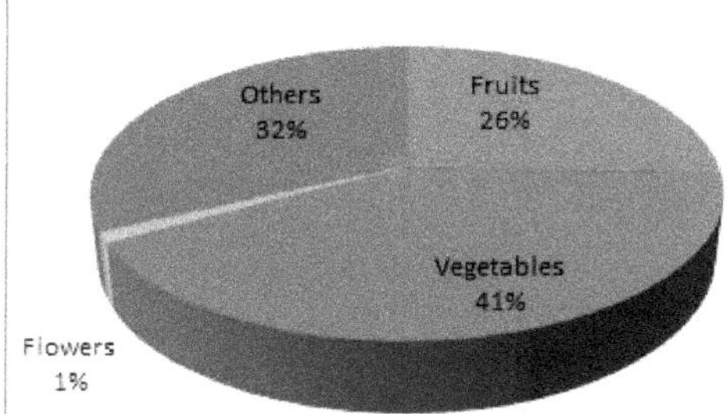

Area and production of vegetables, fruits and flowers in West Bengal vis-a-vis India during last six years are depicted in Table 5.3. Total area under horticulture in West Bengal increases from 1720.2 thousand hectare in 2011-12 to 1838.2 thousand hectare in 2016-17; total production increases from 27029 thousand tonnes to 30008.2 thousand tonnes. The annual average compound growth of area and production of total horticulture crops in West Bengal is estimated as 1.43 per cent for area and 1.28 per cent for production during the last six years. The corresponding horticulture growth rates of area and production are 1.14 per cent and 2.84 per cent respectively in India. Growth of area under different horticulture crops in West Bengal vis-a-vis India is shown in Figure 3 and the growth of production of crops are depicted in Figure 4.

Table 5.3 Recent Growth of Area and Production of Horticulture Crops: West Bengal vis-a-vis India, (2011-12 to 2016-17)

West Bengal		West Bengal						AACG
Crops		2011-12	2012-13	2013-14	2014-15	2015-16	2016-17	(%)
Fruits	A	216.6	220.6	223.5	228.3	249.2	253.0	**3.38**
	P	3055.4	3172.5	2909.7	3313.7	3516.7	3585.3	**3.60**
Vegetables	A	1330.9	1348	1380.3	1387.2	1391.4	1387	**0.88**
	P	23415.7	25466.8	23045	26354.6	22825.4	25505.7	**0.67**
Flowers	A	23.9	24.4	24.9	25.3	25.6	26.0	**1.70**
Loose	P	63.9	65.1	66.5	68.2	69.6	71.3	**2.22**
Cut*	P	25042.1	25429.1	145.2	148.0	197.1	201.6	**13.55**
Total Horticulture	A	1720.2	1742.1	1778.1	1790.5	1836.9	1838.2	1.43
	P	27029.0	29199.5	26678.5	30398.1	27246.2	30008.2	1.28

West Bengal		West Bengal						AACG
Crops		**2011-12**	**2012-13**	**2013-14**	**2014-15**	**2015-16**	**2016-17**	**(%)**
India		India						
Fruits	A	6704.2	6982.0	7216.3	6109.7	6300.7	6373.4	-2.06
	P	76424.2	81285.3	88977.1	86601.7	90183	92918.0	**3.67**
Vegetables	A	8989.5	9205.2	9396.1	9542.2	10106.3	10237.9	**2.74**
	P	156325.5	162186.6	162896.9	169478.2	169063.9	178172.4	**2.37**
Flowers	A	253.7	232.7	255.0	248.5	277.6	306.3	**4.22**
Loose	P	1651.6	1729.2	1754.5	1658.7	1656.2	1699.4	**-0.12**
Cut	P	75066.0	76731.9	542.5	484.2	527.7	692.8	**8.54**
Total Horticulture	A	23242.0	23694.1	24198.5	23410	24471.7	24850.9	1.14
	P	257277.1	268847.5	277352.0	280986.1	286187.7	300643.0	**2.84**

Note: *A= Area in '000 Hectare, P=Production in '000 MT, AACG =annual average compound growth rate (%) during 2011-12 to 2016-17, *2013-14 onward Cut Flower Production is being given in '000 MT as compared to earlier reporting (2011-12 & 2012-13) in Lakh Numbers.*

Source: *National Horticulture Board website.*

Recently, West Bengal has achieved an impressive growth in area and production of fruits. The area under fruits in West Bengal increases monotonically from 216.6 thousand hectare in 2011-12 to 253.0 thousand hectares in 2016-17 with a growth rate of 3.38 per cent per year during 2011-12 to 2016-17, while in all India level there is a negative growth of area under fruits (-2.06 per cent per year) during the same period of time. Among the leading states in respect of fruits, the position of West Bengal is 12th rank in area, 9th rank in production, and 9th rank in land productivity of fruits in 2016-17. Top-3 fruits producing states are Andhra Pradesh, Maharashtra, and Uttar Pradesh.

West Bengal has occupied first position consistently among Indian states in terms of area under vegetables till 2015-16. In 2016-17, West Bengal ranks second next to Uttar Pradesh in terms of both area and production of vegetables. In terms of land productivity of vegetables, the performance of West Bengal is not satisfactory; ranks 10th among the states. High productivity of vegetables states are Tamil Nadu, Andhra Pradesh, Jammu & Kashmir, and Himachal Pradesh. Vegetables production in West Bengal increases from 23.4 million tonnes in 2011-12 to 25.5 million tonnes in 2016-17. The corresponding figures for India are 156.3 million tonnes in 2011-12 and 178.2 million tonnes in 2016-17. West Bengal has contributed a significant share in all India vegetables; on an average, it shares 15% of production and 14 % of area of all India vegetables (Table 5.4). Recent growth rates of area (0.88 %) and production (0.67 %) of vegetables in West Bengal are found to be lower than that of all India growth of area (2.74 %) and production (2.37 %) of vegetables during 2011-12 to (Table 5.3).

Table 5.4 Percentage share of West Bengal in All India Horticulture Area and Production

		West Bengal's Contribution (%) in All India						AACG
Crops		**2011-12**	**2012-13**	**2013-14**	**2014-15**	**2015-16**	**2016-17**	**(%)**
Fruits	A	3.23	3.16	3.10	3.74	3.96	3.97	**5.55**
	P	4.00	3.90	3.27	3.83	3.90	3.86	**-0.07**
Vegetables	A	14.81	14.64	14.69	14.54	13.77	13.55	**-1.81**
	P	14.98	15.70	14.15	15.55	13.50	14.32	**-1.66**
Flowers	A	9.43	10.49	9.74	10.19	9.23	8.50	**-2.42**
Loose	P	3.87	3.77	3.79	4.11	4.20	4.19	**2.35**
Cut*	P	33.36	33.14	26.76	30.57	37.35	29.09	**4.61**
Total Horticulture	A	7.40	7.35	7.35	7.65	7.51	7.40	**0.28**
	P	10.51	10.86	9.62	10.82	9.52	9.98	**-1.51**

Note & Source: *Same as Table 1.*

West Bengal is one of the leading flower producing states in India. It ranks top position in production of cut flowers; more than 30 % share is contributed by West Bengal in India. In terms of production of loose flower West Bengal occupies 3rd position after Tamil Nadu and Karnataka. In terms of area under flowers, West Bengal rank is 5th highest (contributed 8.5 % of total area under flower in India) in 2016-17 and other leading states are Jammu & Kashmir, Karnataka, Tamil Nadu, and Kerala.

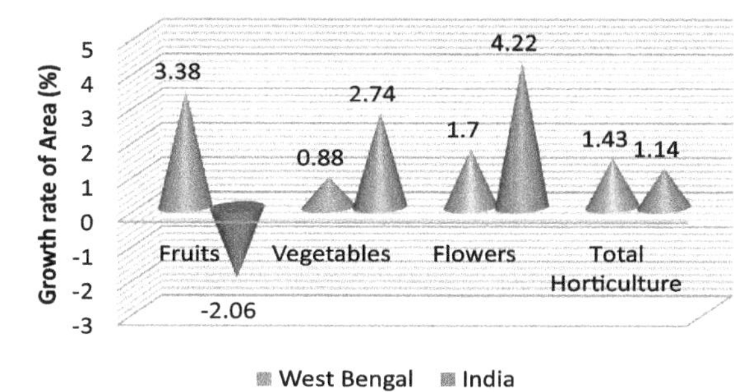

Fig.4: Annual Growth of Area Under Horticulture in West Bengal & India, 2011-12 to 2016-17

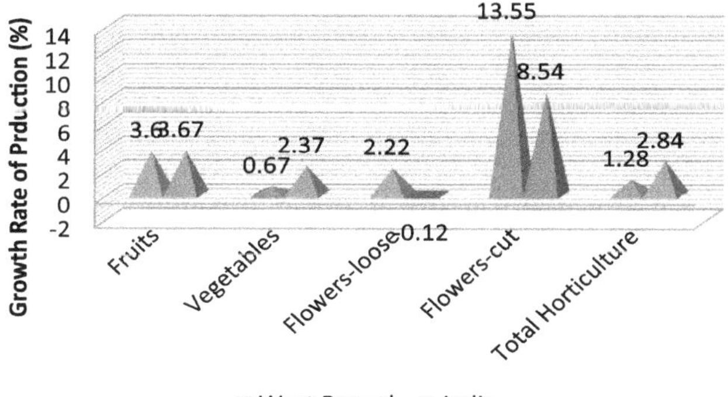

Fig. 5: Annual Growth of Production of Horticulture Crops in West Bengal & India (During 2011-12 to 2016-17)

The area under flowers in West Bengal (India) increases from 23.9 (253.7) thousand hectare in 2011-12 to 26.0 (306.3) thousand hectare in 2016-17; annual growth rate is being 1.70 per cent in West Bengal and 4.22 per cent in India. The contributions of West Bengal in production of both loose flowers and cut flowers are increasing significantly at the rate of 2.35 per cent and 4.61 per cent respectively during 2011-12 to 2016-17 (Table 5.4).

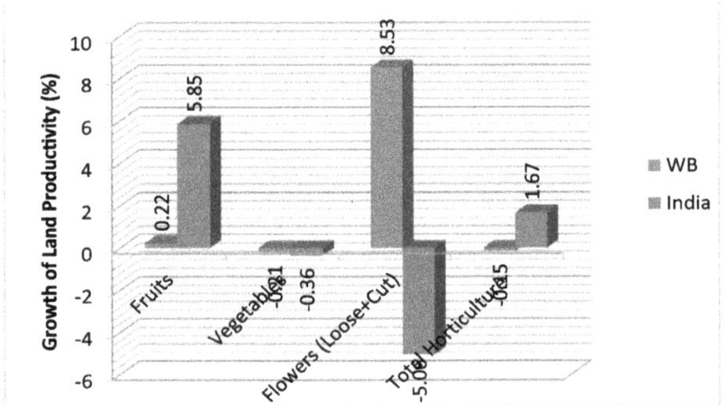

Fig. 6: Annual Growth of Productivity of Horticulture Crops in West Bengal & India (During 2011-12 to 2016-17)

Table 5.5 Productivity (MT/Ha.) of Horticulture Crops in West Bengal & India

Crops	2011-12	2012-13	2013-14	2014-15	2015-16	2016-17	AACG
West Bengal			**West Bengal**				
Fruits	14.1	14.4	13.0	14.5	14.1	14.2	0.22
Vegetables	17.6	18.9	16.7	19.0	16.4	18.4	-0.21
Flowers (Loose+Cut)			8.5	8.5	10.4	10.5	8.53
Total Horticulture	15.7	16.8	15.0	17.0	14.8	16.3	-0.15
India				**India**			
Fruits	11.4	11.6	12.3	14.2	14.3	14.6	5.85
Vegetables	17.4	17.6	17.3	17.8	16.7	17.4	-0.36
Flowers (Loose+Cut)			9.0	8.6	7.9	7.8	-5.06
Total Horticulture	11.1	11.3	11.5	12.0	11.7	12.1	1.67
Productivity Index = (Yield in West Bengal/ Yield in India)*100							
Fruits	123.7	124.1	105.7	102.1	98.6	97.3	
Vegetables	101.1	107.4	96.5	106.7	98.2	105.7	
Flowers (Loose+Cut)			94.4	98.8	131.6	134.6	
Total Horticulture	141.4	148.7	130.4	141.7	126.5	134.7	

Source: *NHB, GOI.*

Table 5.5 depicts land productivity differentials in horticultural crops in West Bengal in relation to all India average productivity. It is observed that land productivity of total horticulture in West Bengal is higher than the National average productivity during 2011-12 to 2016-17; value of productivity index for West Bengal is being greater than 100 in each year. But there are year-wise fluctuations in land productivity both in West Bengal as well as in India. Growth of land productivity is portrayed in Figure 5. As a whole productivity of horticultural crops decreases in West Bengal during 2011-12 to 2016-17; in India it is increasing. Land productivity of fruits and flowers increases and productivity of vegetables decreases in West Bengal. In all India level, fruits productivity increases but productivity of vegetables and flowers decreases during the same period of time.

Education & Earnings

An analysis of aggregated NSSO data (Table 5.6) reveals that average monthly income and consumption expenditure per agricultural households are Rs. 6426 and Rs. 6223 respectively in India during the latest agricultural survey year of 2012-13. The said figures for West Bengal are found to be lower than the national average. In West Bengal, average monthly income and consumption are estimated as Rs. 3980 and Rs. 5888 per agricultural households respectively. Table 5.7 depicts Net attendance ratio for different levels of education by sex in rural West Bengal vis-a vis rural India. It reveals that performance of West Bengal in terms of extent of education (net enrolment ratio and female education) in rural areas is above the national average. Thus, it is very difficult to draw any conclusion regarding the relationship between education and earnings in rural areas based on aggregated data.

Regarding agricultural extension services to farmers it is observed that about 50.6 per cent of the agricultural households accessed technical advices in the field of agriculture from any agencies/ sources in West Bengal, while at the all India level, it is 40.6 per cent of the cultivating households during the period of July, 2012- December, 2012 (Table 5.8). In West Bengal, the major sources accessed by the agricultural households for access to modern technology and technical advices are Progressive farmers (26.4 %), Private Commercial agents (21.3 %), and Radio/TV/newspaper/internet (20.4 %). Only 3 per cent of them reported that they have accessed technical advice from extension agent in West Bengal which is below the national average figure of 6.2 per cent. The contribution of Krishivigyan Kendra in this respect is only 1.5 per cent in West Bengal and 2.7 per cent in India. Private Commercial agents play important role in this regard in West Bengal as compare to all India level. However, farmers' access to modern technology and technical advices in the field of agriculture is a very important aspect for sustainable management of farming practices which positively affects the preferences and practices of the farming community towards better output and income from agricultural activities. As per NSS report more than 90 per cent agricultural households found such technical advices useful in India.

Table 5.6 : Average Monthly Income (Rs.) and Consumption Expenditure (Rs.) Per Agricultural Household During July 2012 - June 2013

Sources of Income	West Bengal	India
Income from wages (Rs.)	2126	2071
Net receipt from cultivation (Rs.)	979	3081
Net receipt from farming of animals (Rs.)	225	763
Net receipt from non-farm business (Rs.)	650	512
Total income (Rs.)	3980	6426

Sources of Income	West Bengal	India
Total consumption expenditure (Rs.)	5888	6223
Net investment in productive asset (Rs.)	147	513

Source: *NSS KI (70/33): Key Indicators of Situation of Agricultural Households in India*

Table 5.7 Net Attendance Ratio for Different Levels of Education by Sex and Residence in West Bengal Vis-a-Vis India

	Rural Areas	
Level of Education	**West Bengal**	**All-India**
Net Attendance Ratio for Male		
Primary	85	84
Upper primary	71	64
Primary & upper primary	90	88
Secondary	49	51
Higher secondary	21	36
Above higher secondary	8	12
Net Attendance Ratio for Female		
Primary	86	82
Upper primary	74	61
Primary & upper primary	90	85
Secondary	55	49
Higher secondary	27	33
Above higher secondary	7	8

Source: *NSS report, 71st Round Survey on Education, NSS KI (71/25.2), June 2015*

Table 5.8 Per 1000 Distribution of Agricultural Households Having Accessed Technical Advice by Source

(during July, 2012 - December, 2012)

Source	West Bengal	Al India
Extension agent	30	62
Krishivigyankendra	15	27
Agricultural university/college	4	12
Private commercial agents (including drilling contractor)	213	74
Progressive farmer	264	200
Radio/tv/newspaper/internet	204	196
Veterinary department	59	80
NGO	6	12
Any agent	506	406

Source: *NSS Report No. 573: Some Aspects of Farming in India, 2012-13.*

We have used NSSO unit (individual) level data to examine the relationship between education and earnings from farming in West Bengal. We have estimated returns to education in rural areas. The Figure 7 shows the estimated average earnings of the rural people engaged in agricultural activities. It varies greatly across districts in West Bengal. It can be said that the income of the agrarian people in these regions differ considerably. The districts which fall under gangetic part are Maldaha, Murshidabad, Nadia, Birbhum, Hooghly, Barddhaman, Howrah, North 24 Pargana and South 24 Pargana. These districts are adjacent to the river Bhagirathi flowing toward the Bay of Bengal. The districts which are less productive and suffer from low productivity of agriculture are: Purulia, Bankura and Paschim Midnapore (belong to Jungle Mahal), Dakshin Dinajpur, Darjeeling, Koochbihar and Jalpaiguri (belong to terai and dooars). The districts with lower productivity have mean earnings below the state average level of income and those with higher average earnings have higher productivity.

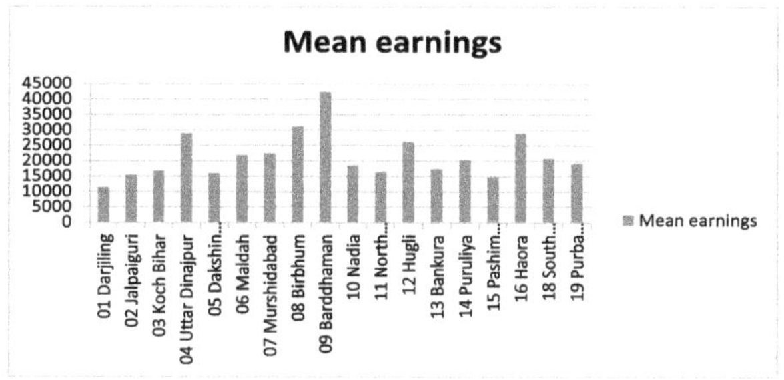

Fig. 7: Mean Earnings of the Rural West Bengal People (District Wise) in Rs.

Source: *NSSO data (SAS, Visit 1, 2012-13)*

Table 5.9 Mean and Standard Deviation (SD) of the Variables Used for Regression Analysis

Variables Used	Mean and (S.D.) (for Total Rural Workers) N= 1609	Mean and (S.D) (for Regular Rural Workers) N=477
Earnings	20997.1 (29992.29)	43502.58 (46223.41)
Years of Schooling	6.13 (5.14)	9.42 (5.33)
Primary	0.28 (0.45)	0.16 (0.37)
Secondary	0.27 (0.44)	0.28 (0.45)
Higher Secondary	0.06 (0.23)	0.13 (0.34)
Graduate	0.07 (0.26)	0.21 (0.41)
Post Graduate	0.02 (0.14)	0.06 (0.24)
Stayed outside	0.12 (0.32)	0.54 (0.22)
Trained	0.02 (0.15)	0.02 (0.16)
Spatial	0.66 (0.47)	0.71 (0.45)
Gender	0.84 (0.36)	0.77 (0.41)
Experience	24.25 (14.22)	22.11 (12.81)
Experience Square	790.53 (805.54)	653.05 (630.42)
Age	36.39 (12.65)	37.54 (11.41)

Calculated on the basis of the unit level NSSO data (2012-13).

It is observed from the analysis of descriptive statistics shown in Table 5.9 that the mean earning is higher (more than double) for those who are regularly employed in comparison with the total rural workers. The average years of schooling is less for the total workers (6.13 years) in comparison with the regular rural workers (9.42 years). The human capital variables used in the model for both the categories interestingly shows that the mean values for the total workers decrease with level of education but the mean values of the regular workers rise with level of education in comparison with the total workers. About 28 % people of the total worker are primary educated whereas about 16 % people of the regular workers are primary educated. If we consider the Secondary education, 27 % people are educated in the total worker category whereas out of the regular workers 28 % people fall in this category. Moreover, for the educational level of Higher Secondary education, only 6 % are in the total workers category but 13 % fall in the regular workers. Similar pattern is observed for Graduate and Post Graduate category. Out of the total workers 7 % are Graduate and 2 % are Post Graduate whereas in the regular workers category 21 % are Graduate and 6 % are Post Graduate. So, it can be said that the regular workers have higher propensity to invest in education and as a result increase the probability of being employed round the year even in the unorganised sector like agricultural economy in West Bengal. The mean value of the persons in the total workers category who have migratory trend is less than the regular worker category but the proportion of people taking training related to agriculture is same for both the categories. More regular workers reside in the gangetic West Bengal than the total workers. The mean value of experience is less for the regular workers in comparison with the total workers but the mean age is less for the total workers in comparison with the regular workers.

Analysis of Regression Results

The table 5.10 shows the OLS estimates of the regression equation mentioned above. The OLS estimation shows that the value of R^2 is higher for the people who were regular wage employed. The first sample consists of the explained variables such as level of schoolings like primary education, secondary education, higher secondary education, graduate, post graduate all are positive and statistically significant at 1 % level of significance for almost all the level of education both in the overall rural people category and the people who were regular wage employed. The difference is that for the later the estimate is robust because it consists of a homogenous group. Moreover, the estimates of the variables like experience and squared experience also show the desired sign. A positive sign on experience implies that income goes up with years of experience which is the proposition that skill increases dexterity and thereby productivity. Moreover, the squared experience negative implies that there is monotony in the same work and thereby productivity goes down. In both the samples the desired signs are observed and are statistically significant. It is observed that the earnings reach maximum when the experience is 31 years. The gender dummy captures the wage differential between male and female which also shows the desired sign i.e. male workers are paid more than the female workers.

Table 5.10 Ordinary Least Square Estimates of the Regression Equation

Dependent Variable: Log Income	All Workers R^2=0.25 N=1609 F=(11,1597)=49.19	Regular Workers R^2=0.34 N=477 F=(11,465)=22.12
Explanatory Variables	Co-efficient (p-value)	Co-efficient (p-value)
Primary	0.105 (-0.100)	0.36 (-0.010)
Secondary	0.391 (0.000)	0.88 (0.000)
Higher Secondary	1.11 (0.000)	1.33 (0.000)
Graduate	1.53 (0.000)	1.63 (0.000)
Post Graduate	2.01 (0.000)	2.01 (0.000)
Experience	0.05 (0.000)	0.05 (-0.004)
Experience^2	-0.0008 (0.000)	-0.0008 (-0.029)
Gender	0.33 (0.000)	0.33 (-0.001)
Trained	-0.15 (-0.320)	0.27 (-0.270)
Stayed outside Native land	0.14 (-0.050)	0.16 (-0.030)
Regional	0.13 (0.000)	0.18 (-0.040)
Cons	8.06 (0.000)	7.758 (0.000)

Calculation is Based on NSSO data 2012-13

In both the samples the results are statistically significant. The dummy variable capturing the effect of obtaining training is negative insignificant indicating that formal training is not sufficient to increase the earnings of the individual. Also, the dummy variable staying outside captures the indirect effect of increasing wages. The indirect effect of staying outside the native land for more than 15 days creates the opportunity for gaining some quality in earning more. However, the direct impacts of basic skills should not be underestimated. Under certain circumstances, just being able to calculate, read and write provides opportunities for the performance of a more lucrative work. However, if the young generation starts to move towards city or migrate towards other areas then the earnings tend to increase which is observed for those who have stayed outside their villages for more than 15 days or like that implying that these labour force has alternative usage to increase the overall productivity of the economy rather than remaining idly in the rural areas. Thus, regression results summarise the following observations:

1. The overall model is significant as the "F" pertaining to each category is significant. The value of R^2 is 0.25 for the total workers and it is 0.34 for the regular workers.
2. The human capital variables are positive and highly significant indicating a direct relationship between education and earnings for both the total workers in the rural area and the regular workers in the rural area. Moreover, as the level of education goes up, the magnitude of the co-efficient also goes up for both the categories though the impact of education on earnings increases with level of education more for the regular workers category.
3. The variable gender is positive and significant indicating that the income of the male workers is upward biased for both the category of workers with same magnitude.
4. The variable for training is not significant for both the category of workers. In case of total workers, the co-efficient is negative. However, for the regular workers it is positive but insignificant. So, the training of the agricultural workers is not up to the mark.
5. The variable representing migratory trend from rural area is positive and highly significant for both the category of workers but the impact of migration increases income for the regular workers more compared to the total workers. In fact, those who get employment opportunity more have a propensity to stay outside the native land in compared to those who are occasional employment.
6. The most important variable is the regional variable in the context of agriculture in West Bengal and it is positive and highly significant for both the category of workers. However, the magnitude for the regular workers residing in gangetic West Bengal is higher in comparison with the total workers.
7. The income of the people belong to rural workers is maximised when the experience is 31 years. This can be calculated by using the formula $-(\mu_1 / 2\mu_2)$ following the study by Agarwal (2011).

The Table 5.11 shows the differential effect of the dummy variables used in the model. According to Halvorsen and Raymond (1980) for the interpretation of dummy variables in a semi-logarithmic equation, the differential effect is captured, since the dependent variable is in the logarithmic form, the coefficient of dummy variable is adjusted by $\{e^{(\text{co-efficient})} - 1\}$. The estimates shown in the table 5.11 suggest that a person (belong to the group total rural worker category) who is primary educated earns 11.07 % more than the illiterate people for the same category. Similarly, with the increase in level of education a person belonging to secondary education earns 47.84 % more than primary educated and so on. A male worker earns 39.09 % more than a female worker. If a worker is staying outside then he is earning 15.02 % more than the person living in the native land. Finally, a person living in the Gangetic West Bengal is getting 13.88 % more than the people living in non-Gangetic West Bengal. The same comparison can be done for the regular workers also. The same trend is observed for the regular workers also but the only difference is that they earn more than the total workers category.

Table 5.11 Differential Effect of Dummy Variables in Semi Logarithmic Regression Equation

Level of Education	Total Rural Workers	Regular Rural Workers
Primary	0.11	0.43
Secondary	0.47	1.41
Higher Secondary	2.03	2.78
Graduate	3.61	4.10
Post Graduate	6.46	6.46
Sex	0.39	0.39
Trained	-0.13	0.30
Stayed outside Native land	0.15	0.17
Regional	0.13	0.19

Calculated on the basis of the estimates of the OLS estimation in Table 5.10

Private Returns to Education

From the OLS estimation mentioned above we will see the private returns to education which is the per year returns to education from individual point of view for investing one more year in education can be calculated by the following formulae Psacharapolous (1994): $r_2 = (\beta_2 - \beta_1)/ n^2$ is the marginal rate of returns to education per year. Here β_2 is the estimate of the co-efficient of 2nd level of education β_1 is the estimate of the co-efficient of the 1st level of education and n^2 is the additional year of schooling for completing level 2. Finally, r_2 is the private returns to education for 2nd level of education. Thus, the marginal returns to education can be calculated using the above formula for each level of education. The table below shows that the private returns to education has increased from primary education to secondary education and from secondary education to higher secondary education in an increasing rate for both the all rural workers (taken together) and for the regular rural workers. But thereafter, the marginal return to education for graduate level has fallen and finally it has increased for post graduate level. The interesting thing is that it has happened for both the type of workers. The only difference is that the returns are higher in lower level of education for the workers who are regular workers as compared with the all rural workers. The returns to higher secondary onwards the returns have drastically fallen for the regular workers in comparison with the all rural workers. The estimates are upward biased for the people in the overall category and even lower for the regular wage salaried people since they get work on regular basis their urge of studying more gets down as they have to forgo the present earnings and there is uncertainty regarding job in the future in the rural area. This also leads to higher drop out ratio for higher education in rural West Bengal.

The marginal returns to education per year of education are higher for regular salaried workers in comparison with all workers for primary and secondary level only. Thereafter with the increase in level of education from Higher Secondary to Post Graduate it is less for the regular workers in rural West Bengal. The possible reason behind this might be those who are regular workers find it difficult to continue education as well as work. Alternatively, it is also true that since the all workers are heterogeneous, outcome changes with level of education. Therefore, as far as policy prescription is concerned, on the job training as done in the organised sectors does not cope up with the same yield in the unorganised sector. So the impact of training on farm income of the people engaged in agriculture is insignificant in rural West Bengal. Moreover, it is observed that only 2.3 % of the regular workers and 2.7 % of the total workers have taken training. Therefore, more and more people have to be trained in the agricultural sector so that with proper training they come under an umbrella of equity.

Table 5.12 Private Rate of Returns to education (%)

Level of Education	All Rural Workers	Regular Rural Workers
Primary	2.63	9.00
Secondary	4.76	8.66
Higher Secondary	35.95	22.50
Graduate	14.00	10.00
Post Graduate	24.00	19.00

Calculated on the basis of the OLS estimation based on NSSO data (2012-13)

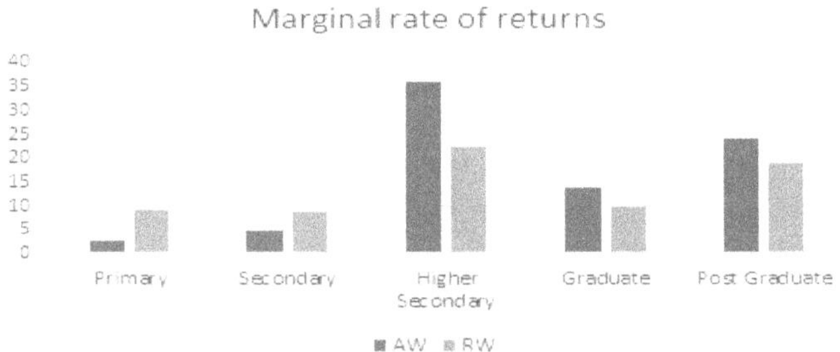

Fig. 8: Marginal Rate of returns to education (%) for the people living in Rural Area.

Calculated on the basis of the OLS estimation based on NSSO data (2012-13)

Conclusions and Policy Prescriptions

Diversification of agriculture towards horticulture is important for sustainable use and management practices of agricultural resources in India. India has occupied 2nd position in the production of world horticulture. In production of some horticultural crops like banana, mango, papaya and Okra the position of Indian horticulture is the highest in the world. Indian horticulture has great potentiality to absorb labour force and to enhance income from farming. Compared to the production of traditional principal crops (paddy- wheat), the production of horticulture crops has been growing faster annually. Moreover, the rate of growth of horticultural products has been accelerated in the post reform era. The contribution of West Bengal horticulture in all India level is very significant. The rate of growth of production is highest for fruits in comparison with the vegetables and flowers in West Bengal. So, diversification of crops in favour of horticultural crops is required in India and for West Bengal; more emphasis should be given towards the production as well as marketing of fruits, vegetables and flowers to ensure management sustainability and profitability of Indian agriculture. Education, training and extension services are the crucial inputs of modern agriculture. Role of education in agricultural sector has significantly positive impact in West Bengal. The study reveals that education has positive impact on the level of earnings from agriculture for both the categories of workers i.e., the total workers and the regular workers. The level of education along with the other explanatory variables explains the Mincerian model more precisely for the homogenous group of workers who are regular. There is wage disparity in favour of male workers as compared with the female workers. So, policies should be taken to empower the female workers for sustainability in agrarian economy. Training in agricultural activities shows desired sign for the homogeneous group i.e., training increases the farm income though it is statistically insignificant. There is regional disparity in the income of agricultural households in West Bengal may be due to diversified agro-climatic conditions and differences on farming practices across districts. People who have migratory trend earn more than others. An adoption of appropriate labour management practices in Indian agriculture is also required. Therefore, as a whole it can be concluded that education and training plays an important role in modern agricultural practices and sustainable management of agri-business. Education planning and manpower planning should be integrated with economic planning to achieve sustainable economic development and management practices in agriculture. The following suggestions are important for increasing employment and sustainable income of Indian farming community.

1. There is an urgent need of educational reforms to change the system of education and training in favour of work conditions and increase productivity.
2. Farmers accessibility to Government Agricultural extension services should be improved for sustainable management and farming practices in India.
3. Farm and non-farm employment opportunities should be created through diversification of agriculture in favour of horticultural crops and marketing management of horticultural products.
4. Emphasis should be given on efficient use of agricultural resources, particularly on land and water.
5. A system should be developed to improve agricultural labour management information for manpower planning and sustainable agricultural development.

References

Adhiguru, P., Birthal, P.S., & Ganesh Kumar, B. (2009). Strengthening Pluralistic Agricultural Information Delivery Systems in India. *Agricultural Economics Research Review*, 22, January-June, 71-79.

Agrawal, T. (2011). Returns to Education in India: Some Evidence, IGIDR. Retrieved from http://www.igidr.ac.in/pdf/publication/WP-2011-017.pdf

Anderson, J. R. (2007). Background Paper for the World Development Report 2008: Agricultural Advisory Services, Agriculture and Rural Development Department, The World Bank, Washington DC, USA.

Babu, S.C., & Joshi, P. K. (2014). Knowledge-Driven Agricultural Development: Policy Options for revitalizing Extension and Advisory Services in India. *International Symposium of International Association of Agricultural Economists (IAAE) on Revisiting Agriculture Policies in the light of Globalisation Experience: The Indian Context. India.* October, 12-13.

Babu, S.C., Glendenning, C.J., Asenso-Okyere, K. & Govindarajan, S.K. (2012). Farmers' Information Needs and Search Behaviours — A Case Study in Tamil Nadu, India. *IFPRI Discussion Paper 01165*, International Food Policy Research Institute. Washington DC, USA.

Becker, Gary S. (1964). *Human Capital: A Theoretical and Empirical Analysis with Special Reference to Education*. University of Chicago Press, Chicago.

Birthal, P.S., Joshi, P.K., Chauhan, S., & Singh, H. (2008). Can Horticulture Revitalise Agricultural Growth? *Indian Journal of Agricultural Economics*, 63(3), July-September.

Duraisamy, P. (2002). Changes in Returns to Education in India, 1983–94: By Gender, Age-Cohort and Location. *Economics of Education Review*, 21(6), 609-22.

Dutta, P. V. (2006). Returns to Education: New Evidence for India, 1983– 1999. *Education Economics*. 14(4), 431-51.

Halvorsen, R. & Raymond, P. (1980). The Interpretation of Dummy Variables in Semi-logarithmic Equations. *American Economic Review*, 70(3), 474-75.

Kingdon, G. G., & Theopold, N. (2008). Do Returns to Education Matter to Schooling Participation? Evidence from India. *Education Economics*, 16(4), 329-350.

Madheswaran, S., & Attewell, P. (2007). Caste Discrimination in the Indian Urban Labour Market: Evidence from the National Sample Survey. *Economic and Political Weekly*, 42(41), 46-53.

Mincer, J. (1974). *Schooling, Experience and Earnings*. New York. National Bureau of Economic Research. Columbia University Press.

Psacharopoulos, G. (1981). Returns to Education: An Updated International Comparison. *Comparative Education*, 17(3), 321-341.

Psacharopoulos, G. (1994). Returns to investment in education: A global update. *World Development*, 22(9), 1325-1343.

Sulaiman, R., & Van den Ban, A. W. (2000). Agricultural Extension in India – The Next Step, *ICAR Policy Brief* (9), http://www.ncap.res.in/upload_files/policy_brief/pb9.pdf

11

Sustainability of Agri-tourism in Punjab: Problems and Prospects

Binali, Chidanand Patil and Chaitra G. B.

Introduction

Tourism is travelling for pleasure or business or any other way also the theory and practice of touring, the activities for attracting, residing and entertainment of tourists, a business for tour operations. Tourism may be international or domestic both. International tourism incoming or outgoing both affects the balance of payments of any country. In today's day, tourism is a major source of income for many countries and has impact on the economy. It accounts 30 percent of world's trade of services and about 6 percent of overall export of goods and services. Tourism also is a good source in generating employment in the service sector of economy associated with tourism in service sector transportation industry benefitted the most like airlines, trains, cruise ships and taxi cabs. Another industry is hospitality services such as accommodations like hotels and resorts and entertainment venues like restaurants, amusement parks, theatres, shopping malls, music venues and casinos etc. world's top tourists destinations includes France, United states, Spain, China, Italy, United Kingdom etc.

Agri-Tourism

Agri-tourism basically rotate around three factors- farm tours, small-scale production of food, and animal husbandry. Guided farm or crop tours fulfil the purpose of both enjoyment as well as education. This rural experience provides an understanding of the farmer's markets, direct sales, and many other features of agriculture that most of us might not even be aware of. From the educational aspect, the tourists who actively involved in Agri-tourism, experience an increasing desire to learn the different ways of growing food, or to learn how animals are raised.

Agri-tourism is a road trip where one can experience and share the rural lifestyle with the other farmers, where the tourist is not merely an observer, but for a while becomes a partner in the different aspects of agriculture. These tourists could be anyone, from a professional employee in an urban area to a businessman or an industrialist.

Agri-tourism is defined as "Travel that combines agricultural or rural settings with products of agricultural operations – all within a tourism experience". Agri-tourism is a dimension of tourism which is economical, as well as enlightening. Besides inculcating the character of rural life in a person, it also engraves some beautiful memories on their minds. The rural life is very simple and ordinary, and Agri-tourism converts this simplicity into an entertaining way of living it. There are a number of words associated with Agri-tourism like Agro-tourism, farm tourism, farm-based tourism and rural tourism. The term used to describe the entertainment aspect of Agri-tourism is "Agritainment". Somewhere or agritainment and agri-education are two sides of the same coin. Agritainment encompasses the aspects of agri-education and vice-versa. (The world of Agri-tourism, 2018). Agri-tourism is the basic idea of bringing urban residents to rural areas for leisure travel and spending. In an increasingly mechanized world, many people have lost touch with how their food is produced, or the region where it originated. Agri-tourism offers tourists a chance to reconnect with the land, providing a "hands on experience" with local foods. Agri-tourism development has manifold benefits directly and indirectly in both ways to the farmers, rural people and tourist also. One of the main benefits to farmers is that it provides income to them even during the offseason, a way to share agricultural passion and to sell the products directly to the visitors and promote the products directly through consumer.

Punjab Scenario of Agri-tourism

Punjab is well known for its culture and heritage not even in India but all around the world. There are many religious places which are very famous among tourists of all around world and tourism in Punjab is religious basically. There are number of Gurudwaras including three principal Takhts of Sikhs, apart from Gurudwaras there are number of hindu temple which are very famous and have history behind them and Muslim places like Rauza Sharif in Sirhind and many other places. So for that Agri-tourism can also become another reason to visit to Punjab by the tourists.

As the name signifies Punjab as land of full bodied five rivers- Sutlej, Beas, Ravi, Jhelum, and Chenab flow through its vast plains located on the north west edge of India as one of the smallest but prosperous state of the country as called as land of people with golden hearts, rich fields and grand heritage, green surroundings and diverse culture, as a perfect location for Agri-tourism. Here all elements exists which keep tourist from across the world enthralled. The state of Punjab has lots of activities and elements to offer to agritourists. Ride on tractor in the fields, relaxing under the shades of tree, taking delicious food of homemade

Punjabi feast, visit to different temples, gurudwaras and mosques with villagers, participating in local games of villages, makes it truly exciting and attractive. To experience the real lifestyle of rural people and experience the diverse cultures of rural Punjab, Agri-tourism is the best source to experience it (Pinky, 2014). For this sake, Punjab government has launched a farm tourism scheme. The Punjab Heritage and Tourism Promotion Board act as a promoter of farm tourism in the form of farm accommodation and activities as the major attraction of India's main agricultural state.

Before the Green revolution in 1960s, there was no such use of technology or advance system in agriculture. Green revolution provides a very rapid forward changes in Indian as well as Asian agriculture sector. Today, due to global trends of concentration, low commodity price, increasing input costs agriculture farmers has to face a lot of competition one another thing is weakened crop growth is also a reason for looking for another supplement sources of income and looking beyond the traditional farming practices. So the farmers are looking for new opportunities to generate the income and Agri-tourism is one main solution for farmers.

So there is large scope and great potential to develop agri-tourism in Punjab as it not only give glimpse of natural beauty but also entertain with the local traditions, customs and foods. Also the cost of food, accommodation, recreation and travel is very less in agri-tourism which broadens the tourist base. Already 31 farmers are presently registered under Punjab Heritage and Tourism Promotion Board. Keeping the potential in the view, present study has taken on the objectives:

Objectives and Methodology

The objectives of this paper are: (i) to ascertain the viewpoints regarding different facets of Agri-tourism, (ii) to study the prospects of Agri-tourism in Punjab, and (iii) To identify the problems faced by farmers engaged in Agri-tourism activities.

The research methodology is the main parameter to judge the worth of any research study. A carefully planned and well documented methodology acts as torch in hand of researcher to carry forward the investigation process.

The study is done at Bathinda area in Punjab. Bathinda is one of the oldest cities in Punjab. It is in the north-western India in the Malwa region and the fifth largest city of Punjab. Bathinda is nick named as the 'City of Lakes', consideration of artificial lakes in the city. The district is situated in the South-western region of the Punjab State and is far away from the Shivalik ranges in the North of the state. It is very near to the Thar Desert of Rajasthan and also far away from the Major rivers lines that run through the state. So, climatically, the district is very hot in summer and frequently blistering heat is in full swing. It has a slight rainy season and a dry but embracing winter. Due to the improvement of irrigation facilities during the past few decades, the weather has undergone a change. Dust storms are a regular feature in summer season when the mercury sometime touches over 47 °C in the summer in June. The monsoon is short and meagre. On the other side the nights as in the desert area are cool and pleasant during December and January & the minimum Temperature at night could touch 0 °C. The soil in the district is mostly sandy. Being sandy Plain region, it is dotted with disseminated sand mounds which have a tendency to shift towards eastern side. But with the development of latest Technology and machinery the topography is under huge change with respect to various aspects connected with green revolution. Now there are various schemes which are functioning to eliminate the spread of desert.

The total of 31 farm houses registered under the Punjab Heritage and Tourism Promotion Board from that list, five agri tourism farms were taken for the survey. Out of which two farms are in Mohali district (Casba farms and Sidhu farms), one is in Roopnagar (Mejie farms), one is on Fazilka-Abohar road (Jyani farms) and one is in Bathinda district (Mann Makhi farms).

As per the requirement of research, the data was collected from the operators of 5 different agri-tourism farms in various places in Punjab Respondents were selected under the Convenient Sampling method. Sample size was 5 farmers, who were selected for the survey considering farmers are already engaged with Agri-tourism activities for more than 5 years.

The questionnaire of farmers dealt about their farm and problems, prospects, reasons to start the farm as tourist place then the suggestions from them. The data were collected personally through questionnaires from five Agri-tourism farms entrepreneurs in Punjab viz., two are from Mohali, one from Roopnagar, one is from Abohar and last one is near Bathinda city. Secondary data were collected through the previous research work done and various journals, articles, newsletters etc.

The data collected from the respondents was transferred on master sheets and data was tabulated and analyzed using the statistical tools such as average, frequency, percentage method.

Results and Discussion

Table 1: Viewpoints of Agri-Tourism Farmers Regarding Different Facets of Agri-Tourism

S.No.	Farm Name	Description
1.	**Casba Farm House:**	The farm house situated in guava orchards near the Mohali district of Punjab. The farm house was constructed at the ancestral land by present owner's father. The farmhouse is being used as farm retreat for the people who are interested in getting a feel of rural Punjab. But right now this farm is under renovation so from past one year no agri-tourism practices are going on there. From Chandigarh airport and railway station it is 20 kilometers away by drive and from bus stand it is 12 kilometers far. The farm house was established in 2009 and the activities provided by them to visit to village lifestyle museum, milking of cows, an exposure of agriculture practices like harvesting of wheat and paddy. The area of the farm is approx. 6000 sq. feet. The owner had received training in agri-tourism.
2.	**Mann Makhi Farm:**	Mann Makhi Farm is situated at village Tungwali of Bathinda district, Punjab. It is 7 km far from Kotfateh railway station and from Tungwali bus stand just drive for 5 km. the basic work done by the farmer is bee keeping. They are very much involved in bee keeping and going to create the cluster of bee keeping at their farm and will provide each and every product extracted from bees. They have gone through many training programs for developing their skills and business. They had started agri-tourism under the governmental scheme named Safurti Scheme. Mostly the school and colleges comes to their farm house to visit and show rural exposure to the students

3.	**Jyani Natural Farm:**	Jyani Natural farm is situated at village Kathera, distt. Firozpur, Punjab. From Bhisiana airport it is 90 kms. The nearest cities to farm are Abohar and Fazilka, away 18 kms from farm. The farm is at 130 acres of land. They do farming on Zero budget or natural farming concept. Harvesting of fruits and vegetables and seasonal crops are done at the farm in all seasons. The activities like swimming, bullock cart riding, entrepreneurial training provided by the farm owners. The visitors are local and outside people, school and colleges and in august to October the foreign visitors use to visit the farm.
4.	**Mejie Farms**	Mejie farms situated in Siswa, Roopnagar. The farm was started by the father of current owner of the farm. This farm is not in Agri-tourism nowadays as after the death of the father of the owner. It is as a subsidiary occupation for them as a tourist place for the people. There are orchards of sapota, Mango and litchi at the farm. They also grow seasonal crops like wheat, paddy on the farm area. Usually, they face the problem of water at their farm for harvesting. They were doing Agri-tourism since 1980 at their farm place. In beginning they were having land of 60 acre but now they have 50 acres of their own.
5.	**Sidhu farm House**	The farm is situated in Majra village near Mohali District of Punjab. The farm house is at the distance of 16 kms from Chandigarh only 30 minutes travel by drive, from Chandigarh airport it is 25 kms away and from Chandigarh railway station and bus stand it has 20 kms distancy. The farm house was established 9 years ago in 2009. The main activities can be performed by tourists in farm are fruits plucking, fishing, walk on special nature trails and farming activities. Bird viewing and flora and fauna are the main attraction at the farm.

From Table 2, it is clear that that all the agri-entrepreneurs were doing the business to increase the awareness about the local products and majority farmers were doing due to their personal interest and saw it as an additional source of income. 3 out of 5 farmers i.e. 60% were innovation seekers and want to recuperate traditional values. Only one person is there who had purpose of changing the traditional farming system. 20 percent farmers were doing agri-tourism by the suggestion of some other person and the same proportion for increasing the long term sustainability. 40 percent wanted to protect the nature and environment. 40 percent were doing to improve the status of their families in society.

Table 2: Agri-entrepreneur's Reason for Adopting Agri-tourism

(N=5)

S. No.	Particulars	Yes	Percentage	No	Percentage
1	Increase long term sustainability	1	20.00	4	80.00
2	Increase local products awareness	5	100.00	0	0.00
3	personal Interest	3	60.00	2	40.00
4	Improve the status	2	40.00	3	60.00
5	Additional Source of income	3	60.00	2	40.00
6	Protect nature and environment	2	40.00	3	60.00
7	Recuperate traditional culture	3	60.00	2	40.00
8	Innovation seeker	3	60.00	2	40.00
9	Innovative way to sell farm products	2	40.00	3	60.00

S. No.	Particulars	Yes	Percentage	No	Percentage
10	Suggested by some other peoples	1	20.00	4	80.00
11	Variation in traditional farming system	1	20.00	4	80.00

Source: Primary data

Prospects of Agri-tourism Farms viewed by Agri-farmers: In the Table-3, W= willing, SW= somewhat Willing and NW= Not willing of farmers regarding the opinions for prospects required, farmers are intended to develop their own website to expand their farm business apart from medical facility to be provided at the farm and for improving the business they are willing to offer real Indian cuisine. To develop the agri-tourism the facilities of medical treatment, having their own website are the facilities on which 80 percent farmers agreed. For the expansion of the area only 40 percent people are willing and 20 percent are somewhat willing. Only one person is interested in incorporating more activities on the farm while two has somewhat willingness and only one person or 20 percent of frequency somewhat willing to provide best accommodation facilities to the farmer but 40 percent are willing and two said they are contented with whatever facilities they are providing as per the charges of farm visit by the tourists.

Contacts with travel agencies is not as much preferred by farmers because they are willing to provide services at their own and reach to the farm by visitors is not the difficult thing to approach so according to them it will not be as much beneficial for them to contact with travel agents. Leading to safety alliance with hospitals and cultural knowledge farmers don't show any willingness because these services are already provided to the tourists on the farm. In advertising activities 40 percent farmers are very much willing to improve their farm visits and 60 percent are willing to contact with schools, colleges, NGOs and other organizations however they are already associated with any or some organizations already.

In Table 4, the problems faced by the farmers in doing agri-tourism business are discussed. The major problems as agreed by all the farmers was that the literature regarding Agri-tourism is not available so people are not aware about the term which creates a big problem for the entrepreneurs and also the lack of funds for publicity agreed by 80 percent farmers, no governmental support said by 80 percent farmers, weak communications skills on the part of the farmers supported by 80 percent farmers creates a big problem to start a tourism business. 80 percent farmers agreed that lack of motivation to farmers to develop a business like that is and lack of training to farmers by all farmers, are also some difficulties faced by farmers while starting the businesses of Agri-tourism.

In the Table 4, the reason for the non-visiting or less tourists visit are discussed with farmers and the major reason come out was lack of awareness according to entrepreneur farmers about 80% farmers agreed with the point whereas climate and lack of tendency are not the big issues in this area as only one farmers told about these issues, only one farmer agreed with these issue whereas 40 percent farmers said that having a farm at very distant place from city also become as a barrier among tourists to visit it.

Table 3: Reviews of Agri-farmers Regarding Prospects of Farm Business

(N=5)

S. No.	Particulars	W	Percentage	SW	Percentage	NW	Percentage
1	To increase or expand the area	2	40.00	1	20.00	2	40.00
2	To incorporate more activities	1	20.00	2	40.00	2	40.00
3	To provide best accommodation facilities	2	40.00	1	20.00	2	40.00
4	To offer real Indian cuisine	4	80.00	0	0.00	1	20.00
5	To make available more agri-products	3	60.00	1	20.00	1	20.00
6	To give better medical facility	4	80.00	1	20.00	0	0.00
7	To contact travel agencies	1	20.00	3	60.00	1	20.00
8	To offer opportunity to participate in rural activities	3	60.00	2	40.00	0	0.00
9	To encourage an offer bullock cart Riding	2	40.00	2	40.00	1	20.00
10	To provide safety with alliance to nearby hospitals	0	0.00	3	60.00	1	20.00
11	To provide information about the culture	0	0.00	3	60.00	3	60.00
12	To arrange programs like folk dance, folk music etc.	2	40.00	2	40.00	1	20.00
13	To advertise more the farm	2	40.00	2	40.00	1	20.00
14	To develop your own website	4	80.00	0	0.00	1	20.00
15	To develop contacts with Schools NGOs etc.	3	60.00	1	20.00	1	20.00

Source: *Primary data*

Table 4: Problems in Agri-tourism

(N = 5)

S. No.	Particulars	Yes	Percentage	No	Percentage
1	Lack of funds for publicity	4	80.00	1	20.00
2	Lack of knowledge and skill to farmers	3	60.00	2	40.00
3	Lack of govt. support	4	80.00	1	20.00
4	Weak communication Skills	4	80.00	1	20.00
5	Lack of commercial approach	4	80.00	1	20.00
6	Less number of visitors	2	40.00	3	60.00
7	Harsh Weather conditions	0	0.00	5	100.00
8	Inadequate price for products	2	40.00	3	60.00
9	Small size of land holding	3	60.00	2	40.00
10	Non availability of literature	5	100.00	0	0.00
11	Non willingness of tourist to purchase farm products	1	20.00	4	80.00
12	Lack of Motivation	4	80.00	1	20.00
13	Lack of awareness of technology	4	80.00	1	20.00
14	Lack of technology	2	40.00	3	60.00
15	Lack of Training	5	100.00	0	0.00
16	Complexity in getting license from govt.	4	80.00	1	20.00

Source: *Primary data*

Table 5: Opinion of Agri-tourism Farmers Regarding Fewer Visits of Customers

S.No	Particulars	Percentage
1	Tough climate	20.00
2	Repellent spots	20.00
3	Lack of tendency	20.00
4	Very distant place from city	40.00
5	Lack of awareness	80.00

Conclusion

Many farmers opined that there is a lack of awareness among the people about Agri-tourism concept as its still in a nascent stage of development. Majority of the Agri-tourism farmers are interested to promote traditional culture by way of Agri-tourism in their area. And to prosper the Agri-tourism business the basic facilities like good Indian village food, better medical facilities and give some glimpse of agriculture and farming to the visitors must be provided agreed by Agri-tourism farmers. The actual problems on ground level which were faced by farmers are like lack of funds for publicity, lack of government support, lack of training among farmers and complex process to be licensed for Agri-tourism. So in a way, we can say, agri-turism is a new approach to tourism but due to some loopholes can't be promoted well. With the government initiatives and farmers support, it can create a lot of scope for tourism.

References

Pinky S. (2014). Agri-tourism in Punjab- A case study, *M. Scthesis. Univ. PAU, Ludhiana, Punjab.*

Persaud. B. (2014). Prospects of Agri-tourism in Ludhiana district of Punjab *PAU, Ludhiana.*

The World of Agri-tourism – *Economics for World Welfare* (2018).

www.punjabtourism.gov.in

12

Natural Resource Management for Environmental Sustainability

S. N. Tripathy

Introduction

Natural resource management consist of a broad spectrum of activities that exclusively require the participation of local communities for instance, micro-watershed management, irrigation water management, soil and water conservation, land and forestry and conservation of biodiversity. Humans have changed ecosystems more rapidly and extensively predominantly to meet the escalating demands for food, fresh water, timber, fibre and fuel. It is estimated that half of the productivity of the Earth's bio-systems has been diverted to human use, with concomitant depletion of our natural resources and impairing the capacity of life-supporting ecosystems (Ehrlich & Ehrlich 1990, World Resources Institute 1998).

The interrelationship between human factors and natural resources management is intricate and has continued at the focal point of the development deliberation. With the rising population and increasing pressure on resources the arguments about 'limits to growth' were raised. Warnings about the world's exhaustible resources such as forests, minerals, petroleum etc., and the need for regulating their exploitation to avert future crises, gave rise to conservationist movements. The environmental degradation, due to ruthless exploitation of natural resources for commercial gains, became a major area of concern. However, the initial concern was more about physical environment (air, water, land pollution and impact on heath etc.) in the early 1970's the focus shifted from 'no-growth' in high consumption countries to more sustainable and environmentally sound development.

It has been reinforced by the concept of 'socio-ecological systems', which recognizes that humans and their social and political systems are an integral part of ecosystems. The successful natural resource/ ecosystem management therefore requires sustainable and equitable governance of the eco-system. It aims to balance human social, economic and cultural needs (the delivery of essential ecosystem services) with ecosystem sustainability (maintaining healthy, productive and resilient ecosystems in the long term). 'Ecosystem Management' is distinct from 'natural resource management', which tends to refer to management of a particular resource (e.g. timber, water) for human use, but it is similar to the more holistic 'integrated natural resource management'.

Natural resources provide food and a wide range of other products, such as fuel, fodder for animals, mahua flower, honey, timber, medical herbs, construction materials, etc. to the mankind. Moreover, natural resources provide services including watersheds, carbon sequestration and soil fertility- on which all human activity depends. Population increase has led to dwindling forest cover, depleting water tables and land fragmentation. Besides, the indiscriminate use of chemical fertilizers and pesticides has led to degradation of soil.

Forests are dwindling due to reckless and indiscriminating use of forest products for paper, lumber and fuel is accelerating the process. Some of the central issues in the debate on forestry policy have been: i) the commercial aspect of forestry operations in India; ii)its implications for forest conservation; iii) the plight of forest dependent communities in India. The dominant viewpoints within the government and forest bureaucracy, continue to blame the population pressure and consequently their dependence on forests to meet their subsistence needs (clearing of forest for cultivation, collection of fuel, fodder and minor forest produce) as the major cause of depletion of the forests. In such appalling conditions like the lack of adequate infrastructure and services, unsafe housing, lack of drinking water, inadequate nutrition and poor health people are adversely impacted. In fact, access to safe shelter, water, sanitation, proper drainage, and reliable solid waste removal, transport, roads and public health services remains a mysterious objective for many of the poor people including India.

In contrast, environmentalists blame the short-sighted policies of the Government which leads to over exploitation of forests for industrial and commercial purposes without an effective program for regeneration of forests and forestry developmental activities (irrigation, communication, and mining). The distorted policy of the Government neglected village commons and grazing lands, forced people to encroach on forests, as the key factors contributing to the denudation of forests.

Objectives and Methodology

In the light of the above-back drop, the paper attempts: (i) to focus on the problems of natural resources management and challenges in its effective implementation for environmental sustainability; (ii) to examine on the environmental challenges of sustainable development issues with particular attention to natural resource management, environment and climate change in the food and agriculture sector; and (iii) to provide a holistic vision for natural resources and the synergies which ensure food security and make agriculture part of the solution to achieve sustainable development.

The methodology adopted in the study is mainly based on secondary sources of data and non-participant observations. Secondary sources of data collected from books, journals, reports, have been used. Many of the issues inter-connecting the problems of environmental and ecological issues and their management from the sustainable development context have been examined and portrayed.

Biodiversity, Conservation and Preservation

Biodiversity has been defined by as "the total variety of life on Earth" which reflects the range of species at a site, the size of the gene pool, or even the number of different ecosystems on the planet. Biodiversity is of paramount significance for sustainable development because it represents the wealth of biological resources available to us and for the future generations in terms of food, clothing, medicine and housing. To safeguard the biodiversity, there is need to reduce natural habitat destruction, undertake national biodiversity assessments, formulate strategies for conservations and sustainable use biological diversity, simultaneously promoting co-operation among countries. Several nations have signed the Convention on Biological Diversity, which forms an international effort to conserve plant and animal species.

Those who are concerned with protecting the environment frequently use the terms conservation and preservation which are though confused but generally mean the same perspective even though there are some differences. Conservation is the sustainable use and management of natural resources including flora and fauna, animals and wild life, water, air, and earth deposits. Natural resources may be renewable or non-renewable.

The conservation of renewable resources like trees involves ensuring that they are not consumed faster than they can be replaced. The conservation of non-renewable resources like fossil fuels involves ensuring that sufficient quantities are maintained for future generations to utilize. Conservation of natural resources usually focuses on the needs and interests of human beings, for example the biological, economic, cultural and recreational values such resources. The rain forest for example, contains a wide range of biodiversity, providing food stocks for local populations and a source of timber and medicines for others.

Preservation attempts to maintain in their present condition areas of the earth that are so far untouched by humans. This is because of the fact that mankind has been encroaching to the environment at a staggering speed. As a result of which many uncultivated landscapes are being diverted for the use as farming, industry, housing, tourism and other human developments. Sustainable development is defined in the Brundtland Commission Report[1] as: "development that meets the needs of the present without compromising the ability of future generations to meet their own needs."

Land and Water Resources

Agriculture occupies a significant position in the Indian economy and provides a source of livelihood for a majority of the population. The imperatives of increasing food grain production to feed India's growing population resulted in increasing attention of agricultural development policies. A major area of concern, however, has been that the growth in agricultural output in the recent past has not been commensurate with the increased inputs (irrigation, fertilizers, seeds, subsidies etc.) and the reasons are multiple.

With the rise in food grain production, the per capita consumption of food grains has not been rising visibly which implies that a more balanced growth in agriculture has not taken place and the purchasing capacity of the rural poor has remained low. Another aspect of imbalance has been the crop-wise disparities in growth between food grains and non-food grains.

Water tables are falling resulting in shrinking of cropland, soil erosion and desertification. Moreover, crop yields are threatened by rising temperatures, global warming and inadequate water supply ultimately leading to drought. The challenges related to management of natural resources and environment is numerous and is exacerbated by climate change which has detrimental impact on agriculture. Climate change is highlighted as a threat to food security, natural resources i.e. land, water, forest resources and biodiversity and coastal resources. The objective of the environment and natural resources management sector is to find out the ways and means to promote sustainable use and management of natural resources and to promote adaptation to climate change through the introduction of climate resistant seeds.

The problem of water scarcity has loomed large due to falling water tables resulting in shrinking of cropland, soil erosion and desertification. Moreover, crop yields are threatened by rising temperatures, global warming and inadequate water supply ultimately leading to situations like drought. Agricultural water use has been the crucial for increasing food production in most of the developing countries including India where water scarce is looming large. To meet the food production for all people productivity of agricultural crops / vegetables has to be increased with optimum use of water; which can be achieved by better agricultural practices like organic farming, fertigation and growing of less water intensive crops.

Conventionally, India has been well gifted with large freshwater reserves, but the increasing population and overexploitation of surface and groundwater over the past decades has resulted in water scarcity in some regions and as such, if adequate and sustainable water management initiatives is not implemented by 2025, India will encounter severe water scarcity. Increased urbanization is driving an increase in per capita water consumption in urban areas with a change in consumption patterns and increased demand for water-intensive agricultural crops and industrial products.

At the global level freshwater reserves are rapidly depleting and this is expected to significantly impact many densely populated areas of the world. Middle income developing regions as well as highly developed countries will encounter water stress in the future, unless existing water reserves are managed effectively. Though low and middle income developing countries currently have low per capita water consumption, but the problem of rapid growth in population coupled with inefficient use of water across sectors is anticipated to lead to a water scarcity in the future. Hence, there is a need to focus on reducing their consumption through improved water management techniques and practices.

Management of water resources for ecological balance is imperative in order to mitigate the detrimental effects of extreme climatic conditions of drought on crops, human and livestock population for their overall improvement (Tripathy, 2015). Secondly, it is essential to restore ecological balance by harnessing and developing natural resources i.e. water, vegetative cover etc. Thirdly, it is needed to boost village community for sustained community action for the operation and maintenance of assets created as well as for further development of the potential of the natural resources in the watershed.

Development planning provides a basis for evolving methods of: i) maximizing human and natural resource development; ii) the investment strategy for utilization and allocation of resources finally and iii) the existing relationship between resources and use rights of people. In rural areas livelihood systems in rural areas are directly dependent on land, water and eco-systems. Transformation from common to private or state maintained resources of these resources also results in transforming the livelihood systems leading to migration; casualization labour; turning skilled artisans into unskilled wage labourers.

In the wake of urban development, industrialization, and development planning there was need for mining and quarrying activities. For agricultural development there was need for big and medium irrigation projects. Thus, all these developmental initiations have taken their tool of India's cultivable lands and forests. The argument here is not anti-development but for a more judicious, holistic and humane development.

Every time a big irrigation project is planned, forest and land laws are enacted or modified and investment policies for use of natural resources are determined. This affects people's relationship to resources and their well-beings. Some of the major environmental problems caused by large irrigation projects discussed by them (Biswas and Biswas, 1984) were: i) increase in water borne diseases, ii) secondary salinization and alkalization of productive land, iii) population displacement, iv) social and economic dislocation effect on people.

The destruction of forests as a common property resource through fencing of forests and consequent changes in the attitudes of policy makers and people to 'forests as a resource', due to changed market conditions, clearly indicates how market demands have dominated over basic needs in our resource utilization strategies.

In a country like India with a population of 130 million having intractable problems of poverty and unemployment ought to make better use of its human and natural resources i.e. land, water and forest. In India about 30 percent the population lives below the poverty line and a segment of the poverty stricken population mainly the tribes and forest dwellers depend for their subsistence on common property resources. Hence, exploitation of land and forest resources increases the degradation of common resources. It goes without saying that the degeneration in common property resources like village woodlots, grazing pastures etc. have resulted in deterioration in living condition of asset-less households.

Moreover, privatization of land, extension of government control over forests and water resources, penetration of market forces are collectively responsible for further aggravating the conditions of poor households. These have hanged the non-monetized economy of the villages which provided for basic needs of the non-land owning households. It is inferred from the above that no other group is more adversely affected by environmental destruction, as the poorer households who are more dependent on the biomass for their life support activities. The changes brought about in livelihood systems through state policy interventions relating to use and management of resource base, may affect poorer household adversely unless they are given viable alternatives. Thus, improving the natural resources base is central to any effort to arrest this "vicious cycle" and increase the productivity of agricultural labourers, who constitute the largest group of people below the poverty line. Ecological degradation, erratic rainfall and a high risk of drought in the area have resulted in food insecurity, increasing out-migration and periodic deaths of the poor people (Tripathy, 2015).

Farmers mostly use chemical fertilizers and pesticides on their agricultural lands in order to increase the productivity so that they can secure a livelihood for themselves. To conserve biodiversity so as to improve the living conditions of these populations, there is a need for an economically sustainable, ecologically suitable production as well as processing and marketing systems. This can be feasible if *sustainable agriculture* - a type of *agriculture* that focuses on producing long-term crops and livestock while having minimal effects on the environment is adopted. *Sustainable agriculture* thus implies to farmers who seek to develop more *sustainable farming* systems. This type of *agriculture* is a productive, competitive and efficient way to produce safe *agricultural* products which tries to find a good balance between the need for food production and the preservation of the ecological system within the environment. Sustainable agriculture conserves land, water, plant and animal genetic resources. Sustainable agriculture is environmentally non-degrading, technically appropriate, economically viable and socially acceptable method of agriculture. Thus, it is inferred from our discussion that efforts to sustainably manage natural resources, increase resilience of livelihoods to threats and crises, are contributing significantly towards sustainable development. All these efforts depend on the management capacity of institutions, on coordination, and on governance mechanisms established to develop and implement judicial, legal and regulatory frameworks. It is highly needed to manage and sustain the eco-systems and natural resources of the community by preventing, arresting and reversing the effects of environmental degradation, through proper management and the sustainable utilization of natural resources.

Integrated Water Resources Management Planning (IWRMP) demonstrates a pragmatic approach in the direction of better and more sustainable water management (UNEP, 2012). IWRMP adopts a process for promoting and coordinating development and management of water, land and interrelated resources, without compromising the sustainability of dynamic ecosystems. The sustainable use of natural resources and the environment with a view to producing goods and services in agriculture, livestock, and forestry depend largely on the way in which individuals, communities and other groups are capable to gain access to land and forests. Accountable governance of tenure of land, and forests are pre-conditions for ensuring social stability. Sustainable use of the environment is imperative leading to responsible investment for sustainable development which in turn paves the way for eradicating of poverty and food insecurity in rural areas.

Degradation of Natural Resources and Climate Change

Climate change is one of the greatest challenges to be tackled; which calls for strong political-will among global leaders. The world leaders assembled at Copenhagen ended the convention with this resolution: "We emphasize our strong political will to urgently combat climate change in accordance with the principle of common but differentiated responsibilities and respective capabilities. To achieve the ultimate objective of the Convention to stabilize greenhouse gas concentration in the atmosphere at a level that would prevent dangerous anthropogenic interference with the climate system, we shall, recognizing the scientific view that the increase in global temperature should be below 2 degrees Celsius, on the basis of equity and in the context of sustainable development, enhance our long-term cooperative action to combat climate change. We recognize the impacts of climate change and ...the need to establish a comprehensive adaptation program including international support," (Draft Decision /CP.15, Copenhagen Accord, 18 December, 2009).

Concerns about climate change, disruption of the global carbon cycle, carbon stocks, and emission and rates of sequestration have, besides supplementing a new dimension to forest management, also transformed forests from a local to a global resource. A new form of economic activity has spawned in the era of global warming, i.e. buying and selling of environmental services. Carbon sinks are created through conserving existing forests and taking up tree-planting projects to remove greenhouse gases.

There is growing consensus today that society is facing a sustainability challenge of a potentially catastrophic nature (Kiron et al. 2012; World watch Institute 2013). Both human and natural systems are being deeply affected by large-scale, systemic challenges such as climate change, species extinction, rainforest depletion and desertification. Across the planet, human caused impacts on global biodiversity and ecosystems are escalating.

During last decades, the use of primary resources has been increased due to the mounting global population, the advancement of economies and changes in the habits of populations from high income and developing countries. Some goods, such as fossil fuels, lands, mines /metals and mineral reserves are frequently limited, while renewable resources such as water and recyclable materials are normally mismanaged. Consequently, the sustainable use of resources is becoming compulsory for the progress of current economies in a practical way, in order to maintain a clean environment for long time.

Impacts of development activities on environment have been illustrated in Table-1. It is found that each development activity has its impact on environment.

Table-1: Some Impacts of Development Activities on Environment

Development Activities	Major Impacts on Environment
Forest clearing and land resettlements	Extinction of rare species of flora and fauna, creation of condition for mosquito breeding leading to infectious diseases such as malaria, dengue etc.
Shifting cultivation in upland agriculture	Soil erosion in upland areas, soil fertility declines due to shorter cultivation cycle, which is practiced due to population pressure, flooding of low land areas. The problems could be resolved by terraced cultivation.
Agro industries	Air pollution due to burning of bagasse as fuel in sugar mills, large amount of highly polluting organic wastes, surface water pollution
Introduction of new varieties of cereals	Reduction of genetic diversity of traditional monoculture resulting in instability, danger of multiplication of local strains of fungus, bacteria or virus on new variety
Use of pesticides	Organism develops resistance and new control methods are needed (e. g. in malaria, widespread use of dieldrin as a prophylactic agent against pests of oil palms made the problem worse), creation of complex and widespread environment problems. The pesticides used in agriculture sometimes go into food chain or in water bodies which may result in harmful health hazards.
Timber extraction	Degrades land, destroys surface soil, and reduces production potential of future forests.
Urbanisation and Water resource projects, e.g. Dam, extensive irrigation industrialization	Concentration of population in urban centre's makes huge demands on production in rural areas and put pressures on land, air and water pollution. Human settlement & resettlement, spread of waterborne diseases, reduction of fisheries, siltation, physical changes e.g. temperature, humidity.

Source: *Compendium of Environment Statistics, India 2010, Central Statistics Office,, Ministry of Statistics and Programme Implementation, Government of India, New Delhi*

Therefore, the recycle of decomposable materials, the addition of renewable energies, the treatment of wastewater for its reprocess, the construction of new facilities adopting green policies by reducing the energy consumption and improving proficiencies, and the reprocess of low-value materials, are all solutions that were suggested for the enhancement of new economies in a sustainable manner and by viable means. On one hand, many efforts were introduced at international and local level for the adoption of these proposals into new management policies.

Conservationists, who advocate the vision represented in the ecological paradigm, are concerned in keeping in equilibrium the infinite cycles of restoration, revival and reproduction that characterize all ecosystems. In this perspective, a crucial factor is the resilience of ecosystems. *Resilience* is *the capacity* of an *ecosystem* to tolerate *disturbance,* their ability to return to equilibrium after having experienced shocks and disturbances. The resilience of ecosystems responds to the need to sustainably use, conserve and *restore ecosystems* and their services. Thus, there is a need for eco-system *regeneration.* Eco-system *regeneration* is "the process of renewal, restoration, and growth that makes genomes, cells, organisms, and *ecosystems* resilient to natural fluctuations or events that cause disturbance or damage".

It is clear that our activities leads to degradation of ecosystems, diminishes *ecosystem* services. But when such *ecosystem thresholds* are breached, there will be undesirable and often *irreversible* change can happen. Therefore, beyond a certain threshold of ecological irreversibility, the damage done to the regenerative functions of the ecosystem can be so great that the ecosystem may never recover from it. For example, rainforest as a natural resources ecosystems, which are enormously complex and host a great variety of species, are threatened by human activities including burning, logging, and clearing for agriculture. These activities can irreversibly destroy rainforest ecosystems, sometimes leaving the soil so depleted that it can support neither forest nor agriculture.

Conclusion and Policy Implications

This paper indicates main areas of concern of poverty such as lack / inadequate access to natural resources (land, forest, and water), over dependence, low productivity due to inadequate use of technology, and deficient infrastructure, unemployment and underemployment in agriculture.

Further, in the light of the aforesaid analysis it is deduced that watershed projects are significant as they are drought- proofing, enhancing crop production, protecting environment and generating employment. It is seen that in the water shed program implemented areas, run- off and soil erosion have been reduced, the capacity for water retaining has enhanced the ground water level, irrigated areas and cropping intensity increased; and finally, the employment of the landless people has hiked.

Deforestation, shifting cultivation, over-grazing and improper cropping of undulating land, plugging of natural drains and other kinds of poor land management have been instrumental for runoff, reduced ground water recharge and water scarcity. Such activities result in the deterioration of soil, loss of valuable nutrients, lower yield, flooding of lower lands , sedimentation of small tanks and wastages of precious water to the ocean. Changes in land use, in particular degradation of soil, loss of vegetation cover and over extraction of ground water are likely to affect water availability through their impact on hydrological parameters determining recharge (Chopra, 2003).

Policy Implications

1. Viewed from the above analysis, we can suggest that a holistic approach to natural resource management is needed, giving adequate attention to gender differences and needs, together with ensuring sustainable use of land, water and forest resources.
2. In the above backdrop, we can suggest that our legal frame should be restructured with the right to life of the common man particularly the poverty –stricken tribes as the central point. First of all, all unconstitutional elements should be identified and eliminated from all the laws particularly those concerning the command over resources, ownership of means of production and existence of labour (Sharma, 2010).
3. Forests have not only been disregarded in economic policy but also have been plundered for short-term gain. Sustainable development of forest resources

and their effective/optimum utilization is crucial for alleviation of rural poverty. Hence, what are needed are a strong institutional framework- and an effective legal and regulatory environment ensuring the rights of specific groups among the tribes and forest dwellers being recognized and protected simultaneously providing opportunities to develop sustainable forest business to these communities (Lahiri and Souparana, 2010).

4. With regard to agricultural sector; water usage efficiency need to be improved in the production of water-intensive crops such as rice, wheat and sugarcane; encourage adoption of techniques such as rain-water harvesting and watershed management in agriculture.
5. Encouraging investment in recycling and treatment of industrial wastewater through regulations and subsidies for water treatment plants; implementing policies to make rain-water harvesting mandatory in cities with new construction projects; propagating efficient water usage practices through community based education programs will go a long way in the water resources management for the environmental sustainability.
6. Participation of different types of stakeholders with a substantial degree of decentralization has been considered indispensable for effective and sustainable management as well as conservation of natural resources. It has been conceived that in view of the diverse costs and benefits involved in the management and utilization of the natural resources; there are implications on the part of the community and nation at large.
7. The natural resources being the drivers of national and regional economic development; strengthening the resilience and sustainable management of biologically significant trans boundary freshwater ecosystems; supporting adaptive capacities and resilience to the negative impacts of climate change; developing and harmonizing standards, framework and regulation on pollution control and waste management; strengthening disaster risk reduction management etc. are the essential steps required so as to ensure sustainability of resource use as well as development.

Notes

[1]The Brundtland Report in recognition of former Norwegian Prime Minister Gro Harlem Brundtland's role as Chair of the World Commission on Environment and Development (WCED), also known as "Our Common Future", was published in 1987 by the United Nations through the Oxford University Press. To assess the progress made on sustainable development on the basis of the Brundtland Report, in June 1992 the largest environmental conference was held, attracting over 30,000 people including more than 100 heads of states at Rio de Janeiro, Brazil known as the Rio Earth Summit. The objectives of the conference were to build upon the hopes and achievements of the Brundtland Report, in order to respond to pressing global environmental problems and to agree major treaties on biodiversity, climate change and forest management. Five separate agreements were made at the Rio Earth Summit. These included: i) The Convention on Biological Diversity; ii) The Framework Convention on Climate change; iii) Principles of Forest Management;

iv) The Rio Declaration on Environment and Development; and v) Agenda 21 (a "blueprint" for sustainable development). Together these agreements covered every aspect of sustainable development deemed to be relevant. These agreements committed creating guidelines for a more sustainable future are still adhered to.

References

Biswas, M.R. and Biswas AK, (1984): *Complementarily between Environment and Development Process*, in Environmental Conservation, Foundation of International Conservation, Switzerland

Chopra, Kanchan, (2003): Sustainable use of water: The next two decades, *Economic and Political Weekly*, 32, 3362

Ehrlich P. R. and Ehrlich A. H (1990): *The population explosion.* New York

Kiron, David, Nina Kruschwitz, and Knut Haanaes (2012): Sustainability Nears a Tipping Point Sustainability Nears a Tipping Point. *MIT Sloan Management Review* 53(2), 69-74

Lahiri, Souparana (2010): From Colonialisation to Commodification: The Saga of India's Forests and its people, *Globalization Governance Grassroots series No. 16*, National Centre for Advocacy Studies (NCAS), Pune

Sharma B.D (2010): *Unbroken History of Broken Promises- Indian State and Tribal People*, Shayog Pustak Kuteer, New Delhi

Tripathy, S. N (2015): Forest Rights Act, Water resources and Ecological Management among the KBK tribes of Orissa, *Jharkhand Journal of Development and Management Studies*, 13 (1)

United Nations *Environment Program (2012): Integrated Water Resources Management Planning Approach for Small Island Developing States*, UNEP, 130

World Resources Institute (1998): *World Resources 1998–99. A guide to the global environment: environmental change and human health.* New York, NY, Oxford University Press

World watch Institute (2013): *State of the World 2013: Is Sustainability Still Possible?* Washington D.C: Island Press

13

Remote Sensing and GIS in Natural Resource Management: A Case of Forest Cover in Punjab

Arshdeep Kaur and Ravindra Singh

Introduction

Natural resources are limited and unequally distributed around the world. In spite of it, ever increasing population leads to overexploitation of the natural resources reservoirs. Activities of economic development and industrialization also contributed in overexploitation of resources. Many consequences are emerged when we are unable to control these problems. Overexploitation of natural resources can help in short term benefits, but optimum use of resources can helps in long term development process. Natural Resource Management becomes important aspect to conserve our resources which can be depleted soon. For example in places where there is less amount of any resource, alternative resources should be used rather than exploit them rapidly and only then they can be used for long period of time. Sustainability doesn't mean to live a uncomfortable life but it means to make people aware about consumption of resources, so there can be optimum use of resources with maximum benefits and waste can be reduced. There are many types of problems related to natural resources management. Water, air and land are basic resources which are freely available and necessary for survival but by contamination and degradation of these resources we are making our future in danger. Forest fire, floods, mining, land degradation, water scarcity, climate change and Sea level rise are some major problems emerging due to inappropriate management of resources. In some areas there is extreme amount of water and flood like situations are emerged, on the other hand with over use of irrigation, people making land barren and dry. Sometimes conflicts also arise for these resources, for example SYL (Sutlej Yamuna link) conflict between Punjab and Haryana states. This can also affect the international relations like water dispute regarding Brahmaputra river between India and China. There is need to create a balance in use of these resources.

There are many different ways to manage the natural resources but for effective planning we should know the present status of resources. Remote sensing can be helpful tool to access the fast and accurate information about these resources. This paper mainly focuses on the role of remote sensing in sustainable natural resources management. The way of field survey is more time taking and expensive in comparison to remote sensing. We can compare the past data with present to see the changing pattern and GIS as mapping tool can also helps in decision making to make resource management plans. RS data provide base for mapping and monitoring of these resources and changes. This up to date data provide base map and thematic maps and also helpful for developing countries. Land use mapping through remote sensing provides valuable information for environment planning and land management (Jusoff and Hassan, 1998).

Objective and Methodology

It is with this backdrop, the objectives of this paper are (i) to study the use of Remote Sensing and GIS techniques in the natural resource management, and (ii) to review literatures on study of forest cover in Punjab with application of Remote Sensing and GIS techniques. In this study the existing literatures has been used as a secondary data source. This is a type of descriptive study in which potential fields of applications of Remote Sensing and GIS are identified for natural resource management.

Natural Resource Management

There is need to understand why natural resource management is essential for sustainable development. Natural resources are badly degrading under the pressure of increasing human needs. According to study of NASA out of 37 major water sources 21 are in severely water stress because of more consumption. Water consumption has been increased twice with increasing population. According to NASA, world war third will be fought on scarcity of water (Gupta, 2016). There is also a water war hypothesis according to which increased water scarcity is enough and sufficient condition for occurrence of war (Engelke, 2016). Water management should be in productive manner which should start from the level of large scaling measures for sustainable watch catching system to prevent the floods and drought like serious problems. There can be cost a lot if water managed badly. So there should be Sustainability focused management. Like water other resources are also in danger. Forests are cutting day by day, with deforestation temperature is also increasing, and global warming and climate change like conditions rise. Deforestation also affect the soil resources and other natural resources. Every resource management is linked with management of other resources also if we use one resource in unmanageable way affect could be also seen on other resources.

Use of Remote Sensing and GIS

Remote sensing and GIS are possibly the useful tools for management of natural resources. With an increase in capabilities of Remote Sensing, GIS is also enhanced as a tool of decision supporting with repeatability, resolution and mapping. The quantity, resolution and type of RS data depend on different purpose. Combination of both RS and GIS allow to for improved management strategies for cross basins (Fisher, 2012 and Rao, 2004). There are also some limitations in RS GIS tool like data availability, and limited spectral, temporal and spatial resolution of satellites. Atmospheric conditions can also affects the accuracy of data. Due to limitation of data it restricts us to study at a level of large-basin scale and limited only upto the water budget portion. With time data reliability of satellite data is getting better so it in future it can be proved more useful.

Land Use Classification: By using remote sensing data land can be classified used for different purposes. Land use maps are valuable sources prepared by remote sensing in management and environment planning. SPOT and Land use Thematic Mapper can be used to identify the real shape and size of features due to high spatial resolution. Trees, grassland, and types of vegetation can also be differentiated. We can overlay different database, like identified resource map on the political boundaries and land ownership for land management decisions. For land classification in large areas of tropics landsat data has been used in 1970s. The landsat data has been used for Langkawi Island, in which paddy field, rubber, mangrove forests, inland forests, water bodies and bare land are easily identified. Landuse changes also detectable to RS by using data of two different time periods (Jusoff and Hassan, 1998).

Ground and Surface Water Management: The ground water research objectives are to see quality of water, groundwater fluctuation and artificial recharge site identification and demarcation of groundwater potential zones. It is difficult to with traditional methods of paper maps etc. and time consuming (Jusoff and Hassan, 1998). Changes in fresh water pattern can be seen through remote sensing. Remote sensing satellite data can be used in different stages of watershed management. Data can be used in stage of survey, management, planning, monitoring and implementation in watershed (Rao and Kumar, 2004). Areas of drought, over-irrigation, and wetlands can be identified through remote sensing and GIS. With the advancement in remote sensing now precipitation, temperature and evaporation is also measurable. For quantitative information of precipitation, geostationary satellite thermal IR data and low orbit microwave imager and sound measurements both can be used (Birman, 2015).

RS GIS data have been applied for two water budgets in Africa (Armanios and Fisher, 2012). Landsat TM data has been used with integrated GIS data to estimate the quality of water in Japan. In Singapore, Landsat 5 TM has been used to know the water quality (Jusoff and Hassan, 1998). Famiglietti, 2018, in his report global map is prepared with the help of satellite data, which shows the changes in distribution of Earth's fresh water since 2000. From which they also analyzed the trends of water storage- surface water, ground water, soil moisture, snow and ice etc. nearly for 15 years. Shift in water security and insecurity can be shown by these maps. Some results are found by this using Remote Sensing and GIS like the Southern half of US becomes drier and Northern half becomes wetter. Using GIS three dozen water insecurity hotspots shown on the basis of danger to regions health, environment and welfare of people.

Coastal Resource Management: The coastal areas are important for socio economic wellbeing of country. At present 53 % of population of USA lives in coastal areas. Due to increasing population in these areas lead to pollution, eutrophication of coasts, contamination of sediments, fish disease etc. Remote sensing with integrating numerical techniques can help to address the problems including invasive species, coastal hazards and fisheries management. MODIS, ENVISAT, Sea Wide Field of View Spectrometer (SeaWiFS), RADARSAT and Landsat satellite data and data of many other satellites can be used. Quantitative information of ocean colour provided by variety of satellites can be applied on transport, algal bloom, and source of pollution and sediment plums. Sea grass meadows also can be identified by RS. For example problem of expansion of non indigenous species grass Phragmites australis along Atlantic coasts are identified by wetland managers and ecologists. Dense roots of these species prevents to other species from growing. The high resolution remote sensing imagery can be useful detection, control and management of this. In the management of coastal reserves of Virginia south-eastern US state, hyper-spectral and high resolution multispectral data has been used to detect and monitor the expansion of Phragmite colonies. There is other problem of coastal hazard like coastal erosion, coastal storms, flooding and hurricanes etc., which also increase potential economic loss. Satellite Remote Sensing and GIS has been used in India, Italy, and Egypt in studies of coastal geomorphology and hazard mapping (Donato and Klemas, 2001).

Forest Management: Due to increasing population slowly area is converted into built up from agriculture, from forests area is converted into agriculture. There is need to optimal use of resources with the increasing demand of human (Prabhbir Singh, 2011). In forestry application of Remote Sensing and GIS can be helpful in many ways, and one of them is application in support of sustainable forest management. Remote Sensing and GIS forest inventories also helps in management of forest. Remote Sensing and GIS can be useful to identify the deforested areas by comparing present and past data. Near infrared and short wave infrared are useful to provide specific information about vegetation in terms of moisture and greenness. NIR is sensitive to detect leaves and shortwave IR sensitive to moisture. NDVI (Normalized Difference Vegetation Index) and EVI (Enhanced Vegetation Index) also can be generated to see, presence of actively growing vegetation at the time of observation. NDVI has been used in many applications like vegetation health, deforestation, land degradation, drought detection and famine early warning system. EVI is sensitive to structural variations of canopy, leaf area, canopy architecture etc. and NDVI is sensitive to chlorophyll (Ceccato et al, 2010). Forest fire, disease and insects are some problems which effects forest health. Combined use of Remote Sensing and GIS can help in monitoring and mitigation of insect damage. Which includes detecting damaged areas, pattern of disturbance, predicting hazard and risks, and providing map based pest management decision supporting tool (Michael A, et al.). Susceptibility of forest fire can be evaluated by using RS and GIS. Another important property of Remote Sensing is estimation of vegetation water content by which can be helpful to understand the burning process of biomass. To retrieve the canopy content of moisture in vegetation NIR and SWIR range of wavelength are necessary (Ceccato et al, 2010). In a study of Sungai Karang and Raja Munda forest reserve areas of Malaysia hotspots of forest fire were identified using NOAA 12, NOAA 16 and AVHRR satellite data of year from 2000 to 2005. Landsat-7 satellite data has been used to identify the factors behind forest fire, like fuel type, soil type, agriculture data and NDVI is also extracted. This information can be used forest fire suppression planning and early warning system. Forest fire hazard areas can be identified and classified by making fire susceptibility maps with RS and GIS geospatial technology (Pradhan and Awang, 2006). An improved discrimination in physiological and forest cover can be seen using hyper-spectral Remote Sensing and forest floor characteristics can be seen through RADAR (Radio Detection and Ranging). LIDAR also can be used in forestry to estimate vertical distribution of forests, height and forest biomass (Ceccato et al, 2010). By using GIS mapping of forest hazard areas like fire affected and deforestation can be done and fire mitigation and aforestation plans can be applied on those areas.

Soil Management: when land is used beyond its capacity, many adverse impacts are emerged like soil degradation and soil erosion. Desertification of an area and soil erosion can be caused by a large rate of deforestation which affects the productivity of land. For the protection of productive land the only way is to conserve soil. RS and GIS are powerful tool for soil erosion estimation because of high resolution and synoptic view instead of traditional inaccurate and time taking methods (Rao and Kumar, 2004). In a study of Phewa watershed in Nepal, erosional hazard mapping has been done by GIS using landcover data. In the study of Mae Klang Watershed in Thailand shown that combination of remote sensing data with other data like with metrological data can provide useful information about risk of soil erosion in an area. In another study conducted in China, combination of Landsat 4 MSS data with field investigation has been used. 38 % area is classified to weak, medium and intensive areas according to investigation. By using image map database including elevation data, geological data, landslides map, Landsat MSS data characteristic features of landslide areas can be extracted. In a study attempted in Japan, landslide danger zone is identified by overlying Digital Elevation Model on Landsat TM data (Jusoff and Hassan, 1998).

Forest Cover in Punjab

Punjab is situated in Northern part of India at longitude 31.14° N and latitude 75.34° E. Himachal is situated in the Eastern side, Jammu Kashmir in Northern, Haryana in South East direction, Rajasthan in Southern direction from Punjab, and Pakistan in the Western direction. There are alluvial fertile plains in the most of the area of Punjab and low Shiwalik foothills in the North Eastern part. Climate in this area is tropical, subtropical monsoon, hot, semiarid with hot summers and cold winters. Punjab's total geographical area is 50,362 sq km. By using satellite data Oct-Dec 2015, it is found that there is 1837 sq Km area which is 3.65 % of geographical area under dense forest cover (India State of Forest Report, 2017). According to Principal Chief conservator of forest of Punjab, at present out of total geographical area of Punjab 6.87 % area is under forest cover, which is high in comparison to 2003 when it was 6.3 % of geographical area (Despite limitations, Punjab has improved its forest cover, 2018).

Punjab is agriculture intensive state, 84 % of total geographical area is under agriculture, so there is less forest covered area in Punjab. In Punjab dense forest are found in some areas of Shiwalik foothills which are in the North Eastern part of the state (Despite limitations, Punjab has improved its forest cover, 2018). While the plains are fertile so there is more agriculture cover can be seen in plain areas. Dense forest cover in Punjab mainly found in the districts of Hoshiarpur, Rupnagar, SAS Nagar, Gurdaspur and Pathankot (ENVIS Centre Punjab, 2011).

In a study conducted by Singh and Shastri (2010) in Rupnagar district of Punjab, Land Use and Land Change (LULC) mapping has been done using satellite data. In this study LISS III data of year 2006 and Landsat data of year 1989 and year 2000 has been used. Unsupervised classification has been done which classify land on the basis of spectral reflectance.

Fig.1 Rupnagar Landsat image year 1989 and Unsupervised classification

Source: *Singh et al. (2010)*

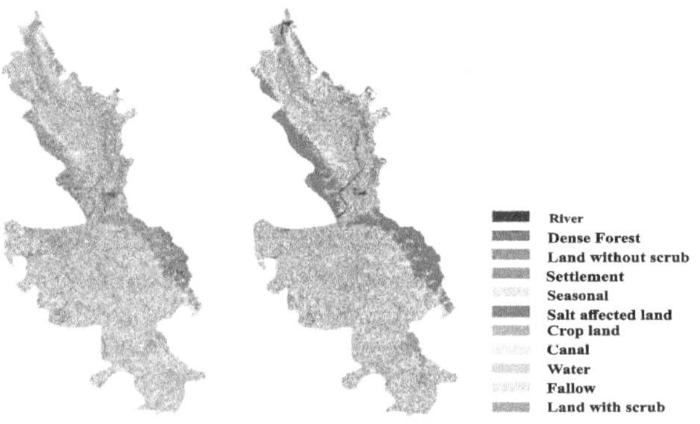

Fig.2 Rupnagar Landsat image year 2000 and unsupervised classification

Source: *Singh et al. (2010)*

Above figures shows the unsupervised classification of Landsat images. Images of two different years with gap are used to show the changes. The images are classified into different classes: Water, Fallow land, Scrub, Canal, Settlement, River, Dense Forest, Seasonal, Salt affected area, Crop land and Water logged area. We can see the changes in these features if we compare both the images. The finding of this study is there is decline in forest cover area from year 1989 to 2000 is about 118.83 Km^2. This can be also correlated to the increased settlement area and fallow land, which can be cause of increasing population. River area is decreased and under the influence of green revolution and over irrigation salt affected area also increased. According to findings of this study due to rapid urbanization there is great decrease in forests land (Chander Kumar Singh, 2010).

In another study of Singh P. and Khanduri, (2011) Remote Sensing has been used to show the Landuse landchange pattern of Pathankot and some areas of Gurdaspur district of Punjab. Satellite images of LISS II, LISS III, ETM+ TM Landsat and IRS 1A, D has been used in this study.

Fig 3

Source: *Singh and Khanduri (2011)*

These figure shows the information of three different time periods of 1991, 2002 and 2006 year. In which major changes in landuse has taken place. Land is classified into 18 different classes on the basis of their spectral reflectance using unsupervised classification. According to Results of this study, there is decrease in area under forests by 47.87 sq /km^2 which is 35.13 % during 15 years from 1991-2016. Area under open forests has also decreased in this period by 30.74 sq/ km. On the other hand built up area has increased from 15.85 sq/km in 1991 to 19.94 sq/km in 2006 in urban areas and 12.7 sq/km in 1991 to 14.65 sq/km in 2006 rural areas. We can see in figure 3, Red colour represents the built-up area, there is large expand in built up area and forest area is represented with dark and light green colour which is diminished. Forest areas are converting into cropland according to study.

In a study of Brar, (2013) conducted on Land use changes in Shiwalik foothill area in Punjab, satellite images of Landsat TM and IRS P6 have been used of year 1989 and 2005. Supervised classification method is used to identify different classify land use in 8 different classes.

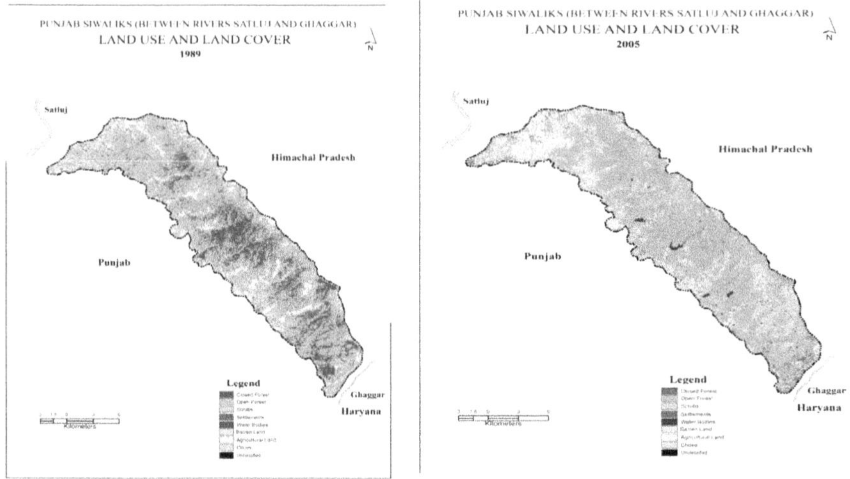

Fig.4

Source: *Brar (2013)*

This are Supervised classified images of landsat of two different years. We can see the difference in land use taken place during this time. There is decrease in area under dense forest cover at very fast rate which is represented with dark green colour and increase in cropland and settlement area. The area under forests in 1989 was 108 sq km and in 2005 area of 38.7 sq km remained under forest cover. Some factors are found in this study, which are responsible for changes in land use. People cutting forests for the fulfilment of their needs of fuels so the area under scrubs is also increasing. These plants are used to feed their animals by people. Barren land is increasing due to deforestation.

Conclusion

Human depends on the natural resources for the fulfilment of his/her all needs. But with extensive use of these resources, most of them will be depleted soon. So there is need to give concern on use of these resources. Natural resources should be used in sustainable way so they can be useful in future also. There are emerging negative impacts on environment with changing land use pattern. All forest areas are converting into crop land and crop land is converting into build up land. Remote Sensing provide up to date spatial data of large area in less time and less cost. Results of these techniques are near reliable Remote sensing can be useful tool for regular monitoring of environmental conditions and rate of changes in those conditions can be reduced. For the restoration of these damaged conditions GIS can help to make future plan maps.

References

Birman, C. F. (2015). Daily Rainfall Detection and Estimation over Land Using Microwave Surface Emissivities. *Journal of Applied Meteorology and Climatology*, 54(4), 880-895.

Brar, G. S. (2013). Detection of land use and land cover change with Remote Sensing and GIS: A case study of Punjab Siwaliks. *International Journal of Geomatics and Geoscience*, 4(2), 296-304

Ceccato, Pietro, and Tufa Dinku. (2010). Introduction to Remote Sensing for Monitoring Rainfall, Temperature, Vegetation and Water Bodies. New York. http://iridl.ldeo.columbia.edu/SOURCES/.NASA/.AVHRR-LST/.LST/.

Donato, T. F., and Klemas, V. V. (2001). Remote Sensing and Modeling Applications for Coastal Resource Management. *Geocarto International*, 16, 25-32.

Engelke, P. (2016, March 22). *Will the world's next wars be fought over water?* Retrieved from Los Angeles Times: http://www.latimes.com/world/global-development/op-ed/la-fg-global-water-oped-story.html

Famiglietti, J. (2018, June 10). *Earth's dismal water future, mapped.* Retrieved from Los Angeles Times: http://www.latimes.com/opinion/op-ed/la-oe-famiglietti-shifting-water-sources-20180610-story.html

Fisher, D. E. (2012). Measuring water availability with limited ground data: assessing the feasibility of an entirely remote-sensing-based hydrologic budget of the Rufiji Basin, Tanzania, using TRMM, GRACE, MODIS, SRB, and AIRS. *Hydrological Processes*, 853–867.

Gupta, A. (2016, May 24). *World War III will be fought over water.* Retrieved from Quartz India: https://qz.com/india/691254/world-war-iii-will-be-fought-over-water/

Jusoff, K. & Hassan, H. M. (1998). An overview of satellite remote sensing for land use planning with special emphasis in Malaysia. *Remote Sensing*, 16(3), 209-231.

MoE (2017). *India State of Forest Report.* Ministry of Environment, Forests and Climate Change.

Pradhan, B., and Awang, M. A. B. (2006). Application of Remote Sensing and GIS for forest fire susceptibility mapping using likelihood ratio model. *Proceedings of Map Malaysia held at Kuala Lumpur, 3-4 May 2006, Available at Geospatial World*

Rao, K. H. (2004). Spatial Decision Support System for Watershed. *Water Resources Management*, 18, 407–423.

Singh, C. K, Shashtri S, and Avtar R. (2010) Monitoring change in land use and land cover in Rupnagar district of Punjab, India using Landsat and IRS LISS III satellite data. *Ecological Question*, 13, 73–79.

Singh, P. And Khanduri, K. (2011). Land use and Land cover change detection through Remote Sensing & GIS Technology: Case study of Pathankot and Dhar Kalan Tehsils, Punjab. *International Journal of Geomatics and Geosciences*, 1(4), 839-846

ToI (2018, June 5). *Despite limitations, Punjab has improved its forest cover'.* Time of India, Retrieved from The Times of India: https://timesofindia.indiatimes.com/city/chandigarh/despite-limitations-punjab-has-improved-its-forest-cover/articleshow/64455830.cms

Wulder, M.A., Hall, R. J. and Franklin, S. E. (2005). Remote sensing and GIS in forestry. *Remote Sensing Applications*, Chapter-12 in Remote sensing for GIS managers, 351-356

14

Sustainable Development Goals and Wetland Sustainability: A Case of Harike Wetland in Punjab

Gaurav Kumar, Amandeep Kaur and Kiran K. Singh

Introduction

Human beings depend over the wetland areas for a wide range of products and services in the form of fishing, water supply, fuel, water purification, spiritual, aesthetic, recreational and tourism (MEA, 2005; Ghermandi *et al.* 2008). In recent times, the increasing demand for water resources and variability in the climate has augmented the significance of wetland. The livelihood of the people living around the wetland area directly or indirectly affected the wetland area as a range of food resources are provided by it. The inland fishes are one of the most significant resources of the wetland area, for which a number of people are dependent for food purposes. The production of inland fish resources has increased from 2 million tonnes in 1950 to 11.6 million tonnes in 2012 (FAO, 2014). On the other hand, wetlands provide area for agricultural activities. About 80 percent people who are living adjacent to the Uganda wetland depend upon wetland for agricultural activities and food (Kakuru *et al.* 2013). Wetland area plays a dynamic role in the regulation of climate through the storage and sequestration of carbon by the vegetation of the wetland area. It is also termed as a carbon sinks (Ramsar Convention on Wetlands, 2018). The role of wetland in climate regulation changed with respect to the presence of vegetation covers and releasing the amount of methane gas. Besides this, continuous evaporation and transpiration from the wetland help to maintain of surrounding climate.

Overall agenda of SDGs revolves around three major values of wetlands; provisioning, regulating and cultural. The fuel, food and fresh water covered under the provisioning services, under cultural include religious and spiritual, recreational, aesthetic and educational. The water purification, climate regulation, flood control and erosion control cover under the regulating services. According to Ramsar Convention secretariat 2018 report entitled 'Global Wetland Outlook: State of the World's wetlands and their services to people' about 75 Sustainable Development Goals (SDGs) indicators directly or indirectly are linked with the wetland area (Ramsar Convention on Wetlands, 2018). The ecosystem services provide by wetland are critical in achieving sustainable development goals (Ramsar Convention on Wetlands, 2018). At the global level, about 1280 million hectare areas are covered by the wetland (MEA, 2005; Finlayson and Spiers, 1999). The significance of wetland areas for the human being changes with respect to wetland types, extent and location (Bassi et al., 2014). Loss or degradation of wetland brings into the force the need for sustainable use of wetlands. As per Ramsar Convention on Wetlands 2018 report, the available data shows that between 1970 and 2015, inland and marine/coastal natural wetlands have declined by approximately 35 per cent. This rate is three times higher than the loss of forests. On the other hand, man-made wetlands in the form of rice/paddy fields and reservoirs, the area has doubled in the same period. Productivity of wetland increased the interest of human being to create more man-made wetlands to ensure livelihood and supply of other services. The regulating services of man-made wetland could not be ignored. This increase of man-made wetland and degradation of natural wetland may not be good for the ecosystem.

Harike Wetland and Sustainability

The Harike International wetland, comprising 4100 hectares area is one of the largest man-made wetland located in the northern region of Punjab State. It was designated in the Ramsar list. Approximately 40500 persons are residing in 7400 rural and urban households within 2 km buffer zone of Harike Wetland (Map 1). People depend upon Harike wetland for water availability, religious and aesthetic values, fishing and fuel collection. With the help of two canals; Rajasthan feeder and Ferozepur feeder, water from wetland is supplied to South-western Punjab and Rajasthan state. Besides, people's dependency on the wetland area is in the form of direct irrigation through shallow dugs of tube wells nearby wetland areas. In ecological terms, wetlands are generally termed as kidneys of the landscapes because wetlands act as natural filters for the purification of water. Water in the wetland areas is treated through presence of hyacinth, as it obtains excessive nutrients from the water for their growth (Department of Forest and Wildlife Preservation (Wildlife), Punjab; Ladhar, 2002).

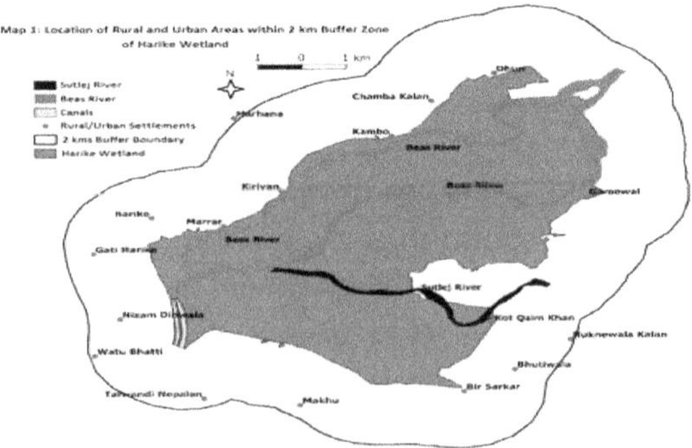

Source: *Map is created using QGIS*

The evidence regarding the acting of natural purifier can be evaluated from the report of Punjab Pollution Control Board which explain about water quality of Sutlej River, which is class 'D' (not fit for drinking with even conventional treatment nor for bathing of human beings and wildlife) before entering the wetland area and class 'B' (fit for drinking without conventional treatment but after disinfection) after the wetland area (PPCB, 2013). In another way, Harike wetland water is useful for agriculture as well as for fishing. It is estimated that 28.65 tonnes/months fish resources extracted from Harike Wildlife Sanctuary in the year 2000 (Department of Forest and Wildlife Preservation (Wildlife), Punjab).

People living around Harike wetland, especially the lower income group, depends on the wetland area for the collection of fuel (Stuip et al., 2002). 'Sarkanda' a type of shrub and other woods found in wetland areas, are used by surrounding people as fuel. Harike wetland is very suitable space for breeding, resting and feeding for different species of flora and fauna. Approximate 100000-120000 migratory birds from various countries like Siberia, Russia, Kazakhstan visit the Harike wetland in the winter seasons every year (Sharma, 2017). Besides this, the wetland area is the centre for the tourist attraction due to its scenic beauty and calmness nature. The area is being used for walking, relaxing, exercising, fishing and other recreational activities. The Harike wetland of Punjab has a great religious importance, as *Gurdwaras* (Sikh Temple) have been situated on both sides of Harike head works. The *Ishardaam* is one of the famous religious places, which is situated on the bank of Harike wetland. The two major festivals are celebrated each year on 13th April and 7th October. Besides this, many religious places are situated on the banks of Sutlej and Beas Rivers. The wetland area also provides livelihood options for women. In the Harike Wetland, a group of 11 women of Sudia village earn their livelihood by the selling of handicraft items by the use of wetland resources.

Harike Wetland and SDG 17

Harike Wetland has potentiality in SDG 17 (Partnership in the Goals) in three different ways; Green Tourism, local Health and Wellbeing, Research/Education. A field survey was conducted in two km buffer zone of wetland to know the potentiality of wetland and its significance for locals. From the field survey of Harike wetland, it was revealed that temperature near the wetland area was low and pleasant than the temperature of far areas. The pleasant temperature, rich flora, scenic beauty of Harike wetland attracts tourists. The potentiality to develop green tourism cannot be ignored.

Green Tourism: According to Ramsar Convention on wetlands (2018), wetland areas are favourite place for relaxing and lessening the tension of life by tourists.. The Harike wetland area has potentiality to attract a large numbers of tourists for recreational, walking, relaxing as well as spiritual need. The concept of green tourism focuses on the sustainable use of the services and functions of the wetland area. There is need to develop the green tourism in Harike wetland which could be beneficial for the livelihood of locals and can conserve the ecosystem services of area. The Harike wetland can be transformed into site of green tourism by adopting environment friendly solutions and restoring them from degradation. Development of certain facilities like bird watching towers, trail road, fishing spots, kayaking, parks and huts houses etc. can attract more tourists.

Health and Wellbeing: The Harike wetland provides a healthy environment for the wellbeing of local people. The report of PPCB found that contaminated water is purified by the wetland area. The water from Harike wetland can be used directly for drinking and bathing purposes of livestock. The dumping of the liquid or solid wastages into the Sutlej and Beas Rivers should have strong policies and regular check. The health of people, livestock, flora & fauna is directly related to the health of water in wetland areas. For the wellbeing of local people, an eco-development scheme around Harike Sanctuary had been developed in 1993-94, in which solar cooker, street lights and fruit trees were distributed by the Department of Forest and Wildlife Preservation (Wildlife), Punjab.

Educational and Research: The presence of a variety of flora and fauna in the wetland area provides an incredible opportunity for research and education. The wetland ecosystem helps to understand how energy flow works or in other words, how energy is transferred from one tropic level to other tropic level. The various types of studies to understand the changing scenario of dependency of livelihoods of people over wetlands can be conducted. It provides a great opportunity for doing scientific as well as social science research. The Harike wetland has a great scope for research in the field of biology, medicine, hydrology, geography, environment and social sciences. At present, education and research is being done on Harike wetland in the form of educational trips, mapping, ecological valuation and study of flora and fauna.

Sustainability at Risk

Presently, Harike wetland area has suffered from various problems like encroachment of wetland area for agricultural actions, dumping of solid and liquid wastages, and excessive growth of unnecessary weeds mainly hyacinth. The encroachment of wetland area or the degradation of the area through the blockage of water in it severely affects the fish production. Such as, according to a study done on Bihar wetland that availability of fish catch has declined from 8-9 kg per day to 400-500 grams due to the decline of the wetland area (Singh, 2018). Around Harike wetland, people residing on the banks of Sutlej River are suffering from the water borne diseases like Malaria and Hepatitis-C. On the west side of the wetland along the Beas River, people do not have such diseases at higher risks. The excess amount of untreated discharge from industries located on Sutlej River through the tributaries mainly 'Chitti Bein' and 'Buddha Nallah' made the water class "D" in comparison to the Beas River where the water is class "B".

Conclusion

At global as well as at local level, wetland plays a key role in the fulfilment of several goals of Sustainable development which are: no poverty, no hunger, healthy life & wellbeing, climate action, life on land, life below water, sustainable cites & production, reduced inequality, decent work & economic growth, clean water & sanitation, quality education and affordable & clean energy etc. The agenda of SDGs is fully or partially supported by the wetland area. But, this cycle of dependency for the fulfilment of the SDG have been affected by over use and degradation of the wetland ecosystem. In present times, wetland are degraded due to land filling, urban expansion, agricultural uses, pollution, or the blockage of the water sources in the wetland area. The sustainable use of wetland products is necessary to enjoy the continuous benefits of wetland area. Because, the unsustainable use of the wetland services and functions can cause severe problems. Therefore, it become necessary to understand the concept of sustainable use and to focus on the implementation of the concept of sustainability in the day to day life.

References

Bassi, N., Kumar, M. D., & Sharma, A. (2014). Status of wetlands in India: A review of extent, ecosystem benefits, threats and management strategies. *Journal of Hydrology: Regional Studies, 2*, 1-19.

FAO. (2014). *The state of world fisheries and aquaculture: opportunities and challenges.* Rome: Food and Agriculture Organization of the United Nations.

Finlayson, C., & Spiers, A. (1999). *Global review of wetland resources and priorities for wetland inventory.* Canberra, Australia: Supervising Scientist.

Ghermandi, A., Bergh, J. C., Brander, L. M., Groot, H. L., & Nunes, P. A. (2008). The economic value of wetland conservation and creation: A Meta Analysis. *FEEM Working Paper No. 79.* Milan, Italy: Fondazione Eni Enrico Mattei.

Kakuru, W., Turyahabwe, N., & Mugisha, J. (2013). Total economic value of wetlands products and services in Ugada. *The Scientific World Journal, 2013*, 1-13.

Ladhar, S. S. (2002). Status of ecological health of wetlands in Punjab, India. *Aquatic Ecosystem Health & Management, 5*(4), 457-465.

Millennium Ecosystem Assessment. (2005). *Ecosystems and human well-being: Wetlands and water synthesis.* Washington, DC: World Resources Institute.

PSCST. (2010). *Annual report 2010-2011.* Chandigarh: Punjab State Council for Science and Technology.

Ramsar Convention on Wetlands. (2018). *Global wetland outlook: State of the world's wetlands.* Gland, Switzerland: Ramsar Convention Secretariat.

Sharma, A. (2017). Smog no bar: Patrolling intensified as migratory birds arrive at Harike wetland in Punjab. *Hindustan Times.* Retrieved from https://www.hindustantimes.com/punjab/smog-no-bar-patrolling-intensified-as-migratory-birds-arrive-at-harike-wetland/story-8DgmS4i7TSGtlNQJIloWyO.html

Singh, G. (2018, October 18). *Bihar's wetlands are on a ventilator, but there is still hope from the ground.* Retrieved from Mongabay: https://india.mongabay.com/2018/10/18/bihars-wetlands-are-on-a-ventilator-but-there-is-still-hope-from-the-ground/

Stuip, M. A., Baker, C. J., & Oosterberg, W. (2002). *The socio-economics of wetlands.* The Netherlands: Wetlands International and RIZA.

Tiwana, N. S., Jerath, N., Saxena, S. K., & Sharma, V. (2008). Conservation of Ramsar sites in Punjab. *Proceedings of Taal 2007: The 12th World Lake Conference* (pp. 1463-1469). Jaipur: International Lake Environment Committee Foundation, Japan.

15

Sustainable Development: A Harmonious Link between the Pillars of Environment and Development

Amit Kashyap, Shashi Bala Kashyap and Syed Wajid-Ul-Zafar

Introduction

"Failure to conserve and use biological diversity in a sustainable manner would result in degrading environments, new and more rampant illnesses, deepening poverty and a continued pattern of inequitable and untenable growth." **-Kofi Annan**

"...Nature provides a free lunch, but only if we control our appetites..."[1]

An appropriate environmental experience has been one of the necessary elements of India's historical philosophy. The civilisation of India has grown up in close association with the nature. There has usually been a compassionate subject for each and every form of existence in the Indian mind. This subject is projected through the doctrine of *Dharma*. The cosmic imaginative and prescient of earth is primarily based on the notion of *'Vasudhaiva Kutumbakam'*. The way ahead will require a flip towards restoration and renewal.

A good environmental sense has been one of the fundamental features of India's ancient philosophy.[2] The civilisation of India has grown up in close association with the nature. There has always been a compassionate concern for every form of life in the Indian mind. This concern is projected through the doctrine of *Dharma.*[3] The cosmic vision of earth is based on the concept of *'Vasudeva Kutumbakam'*. The way forward will require a turn towards restoration and renewal.

Human societies have always relied on biological resources for physical and spiritual sustenance. Biodiversity ultimately provides us with a source of food, medicines, materials and opportunities. The earth's biological resources are vital to human's economic and social development. As a result, there is a growing recognition that biological diversity is a global asset of tremendous value to present and future generations. At the same time, the threat to species and ecosystems has never been as great as it is today. The species extinction caused by human activities continues at an alarming rate,[4] reduction of the earth's biodiversity as a result of human activities is a matter of great concern.[5]

Human activities motivated by the attitude of rampant consumerism and unsustainable patterns of production and consumption have never been as inhumane and callous towards environment as in the modern era of scientific and technological innovations. Man's greed attacks nature environment and ecology and wounded nature backlashes on the human future.[6] We are in the midst of the sixth era of extinction. This problem can be solved only by proper guidance, awareness, education, transfer of advance technology, research, conservation and sustainable use of biological diversity. In order to highlight the importance of biodiversity, 2010 has been selected as the International Year of Biodiversity in an attempt to educate people on biodiversity and how biodiversity supports everyday life.[7]

More than 1200 Summit participants of the United Nations Global Compact (UN Global Compact) from 75 countries met on 21-23 June 2016 in New-York to discuss the issues of contribution of business communities to the realisation of the goals in the field of sustainable development until 2030.[8] The executive director of UN Global Compact Lise Kingo, speaking at the Summit, noted: Many companies start looking at the world through the prism of the goals of sustainable development. Entrepreneurs try to access how their business, goods, and services correspond to the realities of the planet, and how they will meet the demand today and in the long term.[9]

Sustainable development provides a framework for humans to live and prosper in harmony with nature rather than living, as we have done for centuries, at nature's expense. Nonetheless, sustainability does not now have an adequate or supportive legal foundation, in spite of the many environmental and natural resources laws that exist. If we are to make significant progress toward a sustainable society, much less achieve sustainability, we will need to develop and implement laws and legal institutions that do not now exist, or that exist in a much different form. Since their clients in government, business and nongovernmental organizations increasingly demand legal work that addresses sustainable development issues, lawyers have now begun to respond to that demand.

To achieve sustainability, we also need to recognize that, while environmental law is a key to achieving sustainability, it is only part of the necessary legal framework. Other needed legal involve a wide range of other laws, including land use and property laws, tax laws, laws involving our governmental structure, and the like.

This special issue of *Sustainability* is not just about the relationship between law and sustainability. The contributors to this issue also attempt to answer, in various ways, the question of how law can and should be used to achieve sustainability. The phrase, "law *for* sustainability," captures this idea. In a fundamental way, "law for sustainability" is also about governance for sustainability, because law provides essential tools and institutions for governing sustainably. The urgency of the task before us requires this kind of engaged scholarship: providing information, tools, and ideas that policy makers, practicing lawyers, and others can use to address the challenges and opportunities of sustainability.

The contributors to this special issue bring a wide variety of professional, academic, and geographic perspectives to bear on this question. They are widely recognized for their knowledge and expertise.

In this introduction to the special issue, we attempt to synthesize some of the key lessons from the ten substantive articles. Rather than simply summarize each article, we identify and describe their cross-cutting lessons respecting what "law for sustainability" requires. Reading this introduction is thus not a substitute for reading the thoughtful and provocative articles included in this special issue.[10]

Upon creation of the planet, man was rebuked by God to beat it and since then, man has left no stone overturned during this quest to beat the planet. Man is completely hooked into the environment for his daily wants like food, shelter and vesture and it's within the bid to amass these basic wants that he has dealt adversely with the environment. Man's intrusion into this advanced net referred to as environment has had devastating effects. The environment is that the supply of the energy and materials that grouping transforms into product and services to fulfill his wants. It conjointly acts as an enormous sink for the wastes and polluting substances he generates. This increasing hostile and unhealthy environment is inflicting the dislocation, depletion and extinction of species of plants and animals. To keep up the environment at a life sustaining level with attendant economic development and even have a reserve for the longer term, the construct of sustainable development was initiated by the Brundtland Commission.

Sustainable Development is outlined as development that meets the needs of the present without compromising the ability of the future generation to meet their own needs. This principle aims at adaptive the apparent conflicts between environmental protection; economic development and also the quality of life. It's relevancy at the worldwide, national and native levels are its value for environment the context for policy development and environmental laws.

Today, human activities and also the demand for economic process and development have negatively affected the environment thereby inflicting land degradation, pollution, slums within the cities as a results of urbanization, utter neglect and disrespect for the protection of the immediate environment, rather more the future environment. In addressing the issues of economic process, the developed countries have adopted legal ways that decide to meet the competitive demands of urbanization, pollution and also the protection of the environment. The question tackling sustainable development is whether or not non-renewable natural assets ought to be place in peril by economic growth? It's been urged that wherever natural assets area unit in peril, economic process ought to be prohibited to safeguard the environment. This suggestion might control in developed countries however not in developing countries like India wherever economic development is *sin qua non* to human existence. Instead of veto economic development, laws protective the environment ought to be enacted and existing ones invigorated, and compliance ensured to realize environmental sustainable.[11]

Sustainable development may be a flat construct aimed equally environmental element in sustainable consumption of natural resources, protection of environment factors, and health look after population, the social facet by equality, quality of life and stop financial condition, economic by increasing sustainable. Of these area unit equal components of the new development, whose objectives were kicked off on time in documents on the subject sustainable development. This technique consisted of 3 sides: economic, social, environmental, should be considered an entire, and any elements mustn't tend a lot of or less importance than others.

India may be a developing country. A rustic is alleged to be developing once it seeks to advance economically and socially. The event of a rustic principally depends on industries, that area unit the backbone of economy.

For industrial development, we have a tendency to exploit our natural resources indiscriminately. The over exploitation of natural resources and pollution caused by industries leads to environmental degradation. This state of crisis rising from environmental pollution violates the correct to life secure underneath Article 21[12] of the Constitution of India.

Being the soul of basic rights, Article 21 includes the correct to measure in an exceedingly healthy environment, which implies associate degree environment that's free from health hazards arising from environmental pollution. As environment includes water, air, land and also the inter-relationship that exists among and between water, air, land, persons, different living creatures, plants, micro-organisms and property,[13] the adverse impact on any of these elements successively affects the standard of environment.

The right to healthy environment is associate degree inalienable right and at an equivalent time, right to development is additionally imperative. Each human is entitled to economic, social, cultural and political development. Notwithstanding right to healthy environment and right to development area unit historically incomprehensible, each area unit essential for the survival and well-being of grouping. Therefore we'd like a reconciliation construct that never compromises either economic development or the standard of environment. Each development and environment ought to go hand in hand .There mustn't be development at the value of environment and contrariwise. However, there ought to be development whereas taking care and guaranteeing the protection of environment.[14] Sustainable development may be a reconciliation construct between environment and development. It's a technique for continued development while not inflicting damage to the environment. In *Vellore Citizens Welfare Forum v. Union of India*[15], the Supreme Court observed that the normal construct that development and ecology area unit hostile one another isn't any longer acceptable. Sustainable development is that the answer. Sustainable development may be a reconciliation construct between environment and development.

The Notion of Sustainability

Sustainability suggests that the flexibility to keep up a particular state. In ecological context, sustainable is that the ability of associate degree system to keep up ecological method, their functions, biodiversity and productivity for an extended time.[16] Since environment and development area unit different aspects, development should be at intervals environmental sustainable. Meaning we've to bring development while not over exploitation of natural resources.

Environmental sustainable, economic sustainable and socio-political sustainable area unit are the 'three pillars' of sustainable. Merely sustainable is rising the standard of human life whereas living at intervals the carrying capability of supporting eco-systems. The planet and its resources area unit meant not just for the current generation, however conjointly for the generations to return. Therefore, development ought to be at intervals the carrying capability of the environment.

Environment vs. Development

Sustainable development may be a policy for continued development while not touching the standard of environment. It indicates however development ought to be, while not jeopardising environmental interest or to take advantage of environment which isn't confined to the past, present and future generations, in the longer term generation is entitled to the resources which is naturally precocious. Hence, whereas exploiting resources, we should always guarantee their handiness in the longer term to the coming generations. Whereas enjoying correctly the development, we should always conjointly take into consideration of an equivalent right of future generations. For safeguarding the rights of the future generations, the current generation ought to be modest in their exploitation of natural resources. This idea has found widespread international approval since the Maltese Proposal at the international organization General Assembly, in 1967. UN Conference on Human Environment control at national capital from 5-16 June, 1972 was a turning purpose within the history of the construct of sustainable development. National capital Conference has been delineate because the "Magna Carta of our environment".

In the UN international organization Conference on Human environment, the two conflicting concepts; protection of environment and socio economic development were taken into thought. Within the Conference these two contradictory ideas were synchronous by the evolution of a replacement construct that environmental protection is a necessary part of social and economic development. The term 'sustainable development' was coined for the primary time in 1980. It had been within the World Conservation Strategy of the International Union for the Conservation of Nature.[17] World Conservation Strategy self-addressed the truth of resource limitation and also the carrying capability of system. Maintenance of essential ecological processes and equipment systems, preservation of genetic diversity, sustainable utilization of species and ecosystems were the objectives emphasized by World Conservation Strategy.

The World Conservation Strategy was followed by the planet Commission on environment and Development ordinarily called Brundtland Commission.[18] Brundtland Commission in its report "Our Common Future" gave a particular form to the construct of sustainable development in 1987. The commission kicked off the fashionable definition of sustainable development. It defines sustainable development as "development that meets the wants of the current while not compromising the flexibility of the longer term generations to fulfill their own needs". This definition was given by the Norwegian Prime Minister G.H Brundtland WHO was the director of World health Organisation. In 1991 World Conservation Union, UN environment Programme and World Wide Fund for Nature put together made a document referred to as "Caring For the Earth: a technique for sustainable Living" that outlined 'sustainability' as a state which will be maintained indefinitely, whereas development' is outlined because the increasing capability to fulfill human wants and improve the standard of human life. Impressed by Brundtland Report, UN Conference on environment and Development was control in June, 1992 at Rio de Janeiro, Brazil which put the world on the path of sustainable development. Over 150 governments participated during this conference and that was one in every of the biggest gatherings of the governments, heads of state and non-governmental organisations. It's conjointly called Rio Summit or Earth Summit. Rio Summit or Earth Summit is noted because the "Parliament of the Planet". Rio Conference reaffirmed that the implementation of the construct of sustainable development is that the true mode of accomplishment of development.

One of the notable achievements of Rio conference was the sign language of two Conventions. They're UN Framework Convention on temperature change and also the UN Convention on Biological Diversity. The delegates approved 3 non-binding documents that area unit Statement on biological science Principles, The Rio Declaration on environment and Development and Agenda 21- a programme of action. Agenda 21 that is one of the key documents of Earth summit and which suggests variety of measures to uphold the principles of sustainable development. Agenda 21 lays stress on the international cooperation for achieving the goal of sustainable development. It addresses the burning issues of these days and aims to organize the planet to face the challenges of tomorrow.

United Nations Commission on sustainable Development was came upon on sixteenth Feb 1993 in pursuance of article sixty eight of the international organization Charter. This Commission may be a purposeful Commission of the UN Economic and Social Council. The most task of the Commission is to confirm the effective follow up of Rio Conference and enhance international cooperation and rationalising the inter-governmental deciding capability for the mixing of environment and development problems. It examines the progress of the implementation of Agenda 21 at the national, regional and international level.

United Nations organised a 10 days World Summit on Sustainable Development in Rio, African country from twenty sixth August to fourth September, 2002. Rio Conference needed a firm action to resolve the issues known at Earth Summit, 1992 and reaffirmed the member's commitment to sustainable development.

Rio Summit 2012 hosted by Brazil in Rio de Janeiro from 13th to 25th June 2012 was the third international conference on sustainable development. It had been a twenty year follow up to the 1992 earth Summit.[19] It's conjointly called Rio 2012 or Rio+20. The end result of the 2012 Summit was a non-binding document, "The Future we have a tendency to want" and it reaffirmed the Rio principles and action plans. It reaffirmed the thought that economic development and conservation of environment should go hand in hand and bestowed a framework for the creation of an inexperienced economy within the context of sustainable development and financial condition destruction.

Principles of Sustainable Development

Sustainable development may be a reconciliation construct between environment and development. By adopting the principles of sustainable development, the current generation will satisfy their wants while not touching the supply of resources. Therefore, the well-being of the generations to return can never be in peril. In *Vellore Citizens Welfare Forum* v. *Union of India (Tamil Nadu Tanneries Case)*[20], the Supreme Court summed up the principles of sustainable development. They are:

- Intergenerational equity
- Use and conservation of natural resources
- Environmental protection
- Preventive principle
- Defiler pays principle
- Obligation to help and collaborate
- Destruction of financial condition
- Monetary help to developing countries

Intergenerational Equity- The natural resources area unit permanent assets of grouping and don't seem to be meant to be exhausted in one generation.[21] The principle of intergenerational equity lays stress on the correct of every generation of persons to learn from the cultural and natural inheritance of its past generations. Intergenerational equity is that the concept future generations should have an equivalent access to natural resources because the gift generation. The current generation transmitted earth from their ancestors, in order that they have associate degree obligation to pass it on to subsequent generation with an equivalent quality. In *State of Himachal Pradesh* v. *Ganesh Wood Products*[22], the Supreme Court expressed that "the gift generation has no right to peril the security and Well-being of subsequent generation or the generations to return thenceforth." The Indian definition for intergenerational equity is that the "concern for the generations to come".[23] Principle 1 and principle 2[24] of Stockholm Declaration refers the principle of intergenerational equity.

In *Kinkri Devi* v. *State of HP*[25], a PIL was filed alleging that the pseudoscientific and uncontrolled production of the lime stone has caused harm to the Shivalik Hills and was move danger to the ecology, environment and inhabitants of the realm. The Himachal Pradesh court observed that if a simply balance isn't affected between development and environment by correct sound of the natural resources, there'll be violation of Articles 14, 21, 48 -A and 51A (g). The Court went on to state that natural resources have gotten to be tapped for the aim of social development. However, the sound should be through with care in order that ecology and environment might not be affected in any serious manner. The natural resources area unit permanent assets of grouping and don't seem to be meant to be exhausted in one generation. During this case, the court issued associate degree interim direction to the regime to line up a committee to look at the difficulty of correct granting of mining lease and also the necessity of granting lease keeping in sight of the protection of environment.

In *K. Guru Prasad Rao* v. *State of Karnataka*[26], the Court explained the range and scope of intergenerational equity and sustainable development. During this case the appellant filed a PIL praying for the cancellation of a mining lease granted to the respondent and to prevent mining at intervals the radius of 1 km. from Jambunatheswara Temple. The Court control that sustainable development includes preservation and protection of historical/archaeological monumental wealth for future generations. Right to development includes the correct to whole spectrum of civil, cultural, economic, political and human process for the advance of people's well-being and realisation of their full potential.

In *Court on Its Own Motion* v. *Union of India*[27], the Supreme Court control that intergenerational equity may be a part of Article 21 of the Constitution of India.

Use and Conservation of Natural Resources the employment of natural resources should be in an exceedingly sustainable manner. It doesn't need that the whole resources should be reserved for future generations. However, the resources needed for economic process ought to be exploited to the minimum. The thought that, for the advantage of future generations, gift generation ought to be modest in their exploitation of natural resources that has found widespread international approval since the Maltese Proposal at the international organization General assembly of 1967. To realize sustainable development and a dignified of life for all folks, states ought to scale back and eliminate unsustainable pattern of production and consumption. So use and conservation of natural resources is a necessary principle of sustainable development.[28]

Environment Protection– The construct of sustainable development will never be achieved while not protective environment. Development is not possible while not protection of environment and contrariwise. Typically the ways for the protection of environment boost biological process activities. In India, The Environment (Protection) Act, 1986 is enacted for the protection of environment. To implement the selections created at the international organization Conference on Human environment, guaranteeing sustainable development etc. area unit a number of the objectives of Environment(Protection) Act, 1986. Since environment and development area unit complimentary to every different, improvement of 1 isn't doable while not paying due attention to the opposite.

In *Citizen, consumer and Civic Action Group* v. *Union of India*[29] the Court control that there ought to be a correct balance between protection of environment and development activities that area unit essential for progress.

There is no dispute that the society should prosper, however it shall not be at the expense of environment. Within the like vein, the environment ought to be protected, however not at the value of development of the society. Each development and environment shall co-exist and go hand-in-hand. Thus, a balance should be affected and body action need to proceed in accordance there with, and not de-hors an equivalent.

The Precautionary Principle: 'Prevention is better than cure'. The protection of environment will effectively be done by taking adequate precautions against environmental harm. Preventive principle provides stress upon the preventive facet of environmental protection. Preventive principle states that any substance or activity move a threat to the environment is to be prevented from adversely touching the environment, notwithstanding there's no conclusive scientific proof of linking that individual substance or activity to environmental harm.[30] Here substance and activity denote substances and activities introduced as a results of human intervention. Within the context of municipal law, preventive principle suggests that –

- Environmental measures by the regime and also the statutory authorities should anticipate, stop and attack the causes of environmental degradation
- Wherever there are a unit threats of significant and irreparable harm, lack of scientific certainty mustn't be used as a reason for suspending live to forestall environmental degradation
- The headache of proof is on the actor/developers/industrialists to indicate that his action is environmentally benign.[31]

In short preventive principle states that any substance or activity inflicting a threat to the environment is to be prevented from adversely touching the environment. The govt. ought to adopt such measures that anticipate, stop and attack the causes of environmental degradation. If there are a unit threats of significant and irreparable harm to the environment, the state ought to adopt measures to forestall environmental degradation albeit there's no scientific certainty. The damage is prevented even on an inexpensive suspicion. Here the burden of proof lies on the actor to indicate that his act is environmentally sound.

In order to safeguard the environment; the preventive approach shall be wide applied by states in step with their capabilities. Wherever there are a unit threats of significant or irreversible harm, lack of full scientific certainty shall not be used as a reason for proposing price effective measures to forestall environmental degradation. Preventive principle has been recognised in the majority international documents.[32]

In *Andhra Pradesh Pollution Control Board* v. *M.V Nayudu*[33], the Supreme Court copied the origin of preventive principle. The Court discovered that, in initial days the construct was supported 'assimilative capacity' that was advanced by Principle 6[34] of the national capital Declaration of the UN Conference on Human environment.

In *M.C Mehta* v. *Union of India*[35], the Supreme Court applied preventive principle. During this case a PIL was filed alleging that, because of the employment of coal/coke by industries set at intervals the Taj Trapezium Zone, there's environmental pollution and important degradation of Taj Mahal. The Supreme Court applied preventive principle during this case and directed that, all industries in operation within the Taj Trapezium Zone should use gas as a substitute for coal/coke as industrial fuel. The industries that don't seem to be in an exceedingly position to get gas association should stop functioning within the Taj Trapezium Zone and that they ought to shift their industries to different industrial space.

In analysis foundation for *Research foundation for Science, Technology and Natural Resources Policy* v. *Union of India and* Another[36] the Supreme Court created it clear that the "Precautionary Principle" typically describes associate degree approach to the protection of environment or human health supported precaution even wherever there's no clear proof of damage or risk of damage from an activity or substance. It's a vicinity of the principle of sustainable development. It provides for taking protection against specific environmental hazards by avoiding or reducing environmental risks before specific harms area unit full-fledged.

Polluter Pays Principle- defiler Pays Principle is one in every of the fundamental principles of sustainable development. The supreme court of India taken preventive principle within the case *Vellore Citizens Welfare forum v. Union of India.*[37] Defiler pays Principle states that absolutely the liability for damage to the environment extends not solely to compensate the victims of pollution, however conjointly the value of restoring the environmental degradation. In step with this principle, the responsibility to negate environmental harm is upon the defiler.

What defiler Pays Principle envisages is that there's associate degree absolute liability for damage to the environment. The one who is to blame for environmental pollution is at risk of pay compensation to the victims of pollution. The liability of the defiler extends connected the value of restoration of environmental degradation.

According to this principle, the responsibility to negate the environmental harm is upon the defiler. Underneath this principle, it's not the role of the govt. to fulfill the prices concerned in either bar of such harm, or in finishing up remedial action, as a result of the impact of this may be to shift the monetary burden of the pollution incident to the tax payer.[38]

Polluter Pays Principle has been incorporated in essence sixteen of Rio Declaration that provides that national authorities ought to endeavour to push the group action of environmental prices and also the use of economic instruments, taking under consideration that the defiler ought to, in essence, bear the value of pollution, with due relevance the general public interest and while not distorting international trade and investment.

In *Indian Council for Enviro Legal Action* v. *Union of India*[39], a PIL was filed alleging environmental pollution caused by personal industrial units. The economic units situated in Bichhri in Udaipur, Rajasthan were manufacturing bound chemicals like oleum and H-acid while not getting necessary clearance. They need not put in any instrumentality for treatment of extremely venomous effluents discharged by them. These venomous substances percolated deep into the bowels of the planet polluting the bottom water and creating it unfit for drinking by persons and cows and for irrigating the land. The soil became unfit for cultivation. The Supreme Court directed the closure of all such industries. The Court additionally directed the central government to see the quantity needed for finishing up remedial measures together with the removal of sludge from the sites of the industries and also the same shall be paid by the respondent industries. The villagers might claim damages for the loss suffered by them by instituting applicable suits. During this case the Supreme Court applied defiler Pays Principle.

In analysis foundation for *Research foundation for Science, Technology and Natural Resources Policy* v. *Union of India and Another*[40], the Supreme Court elucidated that the "Polluter Pays Principle" essentially implies that the producer of products or different things ought to be to blame for the value of preventing or addressing any pollution that the method causes. This includes environmental price moreover as direct price to the folks or sustainable; it conjointly covers price incurred in avoiding pollution and not simply those associated with remedying any harm. It'll embrace full environmental price and not simply those that area unit forthwith tangible. But this principle doesn't mean that the defiler will soil and acquire it.

The character and extend of price and also the circumstances within which the principle can disagree from case to case.

In *Sterlite Industries (India) Ltd.* v. *Union of India*[41], the appellant company operated the plant while not renewal of licence and did not maintain emission and effluent standards, which resulted in air and pollution. During this case, the Supreme Court applied defiler Pays Principle. Considering the magnitude, prosperity and capability of the corporate, the court directed it to pay compensation of Rs.100 crores.

Obligation to Help And be Co-operative-The construct of sustainable development is achieved solely through international co-operation. Environmental pollution may be a matter of worldwide concern. Therefore it is handled effectively solely with the co-operation of all. Principle 9[42], Principle 10[43], Principle 12[44] and Principle 27[45] of Rio Declaration emphasises the duty to help and co-operate for achieving the goal of sustainable development.

Eradication of Poverty– we cannot succeed the goal of sustainable development while not destruction of financial condition. The construct of sustainable development includes environmental sustainable, economic sustainable and socio-political sustainable. Environment, economy and society area unit the 3 pillars of sustainable. As these 3 area unit mutualist, environmental sustainable will ne'er be earned while not social and economic sustainable. Environmental and economic sustainable results in social sustainable. Brundtland Report observed that financial condition reduces people's capability to use resources in an exceedingly sustainable manner.[46] Therefore, destruction of financial condition is one in every of the principles of sustainable development.

Financial Help to Developing Countries in comparison to developed countries, the over exploitation of natural resources is higher in developing countries. Because the developing countries don't seem to be equipped with adequate technology and monetary resources to tackle the difficulty of over exploitation of resources, the goal of sustainable development is achieved solely through the transfer of technology and monetary resources. Since the developed countries area unit self-sustaining to satisfy their monetary and technological wants, they need to supply their serving to hands to developing countries to uplift them to the trail of sustainable development. International co-operation is required for achieving the construct of sustainable development.

Sustainable Development and Indian Constitution

Sustainable development seeks to keep up a balance between the protection of environment and biological process activities. It's a technique that suggests the manner within which biological process activities area unit to be carried on. For sustainable development, the protection of environment is crucial. Initially, the constitution of India failed to contain any direct provision for the protection and improvement of environment. But the preamble of the constitution provides that India may be a socialist[47]country. In an exceedingly socialist country the state is underneath a requirement to listen to social problems. In *Calcutta Youth Forum* v. *State of West Bengal*[48], the court emphasized that the matter of environmental degradation may be a social problem and underneath the constitution of India the state is duty-bound to require this issue with serious concern. In 1972, the then Prime Minister of India, Mrs. Gandhi created a historic illustration within the UNCHE (Stockholm Conference). In lightweight of India's international obligations arising from the national capital Conference, the ordinal modification to the Indian Constitution in 1976 introduced specific provisions for the protection and improvement of environment. The Constitution (Forty Second Amendment) Act, 1976 obligatory associate degree obligation on the state moreover because the subject of India to safeguard the environment. Article 48-A was inserted as a part of Directive Principles of State policy and Article 51-A (g) was inserted underneath basic duties. Currently the state moreover because the subject is underneath a requirement to safeguard and improve the environment. Therefore, the state shall endeavour shield and improve the environment and to safeguard the forests and life of the country.[49]

In *L.K. Koolwal* v. *State of Rajasthan*[50], the municipal authority underneath The Rajasthan Municipal Authority Act, 1959 was charged with the first duty to scrub streets, removing of baneful substances and vegetation's and every one public nuisances, take away rubbish, odour etc. however the municipality did not perform its primary duty. During this case Mr. Koolwal captive to the court underneath Article 226 and highlighted that the municipality has did not discharge its primary duty that resulted in acute sanitation drawback in Jaipur. The court allowed the official document petition and control that insanitation results in slow poisoning and adversely have an effect on the lifetime of voters and therefore it falls at intervals the compass of Article 21 of the constitution. The Court directed the municipality to get rid of dirt from the town at intervals an amount of six months.

The constitution of India conjointly provides that each subject of India is underneath a requirement to safeguard and improve the natural environment together with forests, lakes, rivers and life, and to own compassion for living creatures[51].

In T. *Damodhar Rao* v. *S.O. Municipal Corporation, Hyderabad*[52], the court observed that in sight of Articles 48-A and 51A(g), it's clear that protection of environment isn't solely the duty of each subject, however it's conjointly the duty of the State and every one different state organs together with courts.

In *M.C Mehta* v. *Union of India*[53] forty two, the Court discovered that articles 39(e), 47 and 48 - A by themselves and together solid a requirement on the state to secure the health of the folks, improve public health and shield and improve the environment.

In *N.D Jayal v. Union of India*[54] forty three, the Supreme Court control that sustainable development is to be treated as associate degree integral a part of life underneath Article 21 of the Constitution of India. Therefore, yielding with the principle of sustainable development may be a constitutional mandate. The Indian Judiciary contend a significant role in reconciliation the 2 apparently opposite concepts; environment and development.

Rural Litigation and Entitlement Kendra, Dehradun v. *Union of* India[55], was the primary case involving the problems with reference to environment and ecological balance that brought into sharp focus the conflict between development and conservation. During this case, the indiscriminate mining within the Mussoorie hills and Dehradun belt denudate the Mussoorie hills of trees and forest cowl. It accelerated eating away leading to landslides and blockage of underground water that fed several streams and comes within the river depression. The court appointed associate degree professional committee to recommendation the Bench on technical problems and on the premise of the report of the committee, the Court ordered the closure of range of lime stone quarries.

In *T.N Godavarman Thirumulpad* v. *Union of India*[56], the Supreme Court prohibited mining activity in Aravalli hills, particularly therein components that has been considered forest space or protected underneath Environment (Protection) Act, 1986.

Conclusion

Development is rarely disjunctive to environment. Development and environment is just like the 2 sides of a coin. They are complimentary to each other and they don't seem to be divisible. Right to healthy environment and right to development is the basic human rights. We have a tendency to have development while protective to environment and contrariwise. Therefore for enjoying the right to development and right to healthy environment at the same time. Like a policy which harmonises these two contradictory ideas that brings development, while not jeopardising environmental interests. Implementing the principle of sustainable Development is the best manner of achieving development while not inflicting adverse impact on the environment.

For attaining the goal of sustainable development, a modification in angle is needed. The consumption of resources ought to be proportionate to the supply. Biological process activities ought to be at intervals for the carrying capability of the system. 3R approach (reduction, recycle and reuse) may be a principle/goal to succeed sustainable development. Sustainable development may be a harmonious construction between environment and development.

This is a cynical moment in earth's history, when humanity must choose it's time for the future. Our planet earth is perhaps the only human habitat in the vast universe and we owe it to posterity to preserve the divine heritage of our biosphere without pollution, degradation and destruction. The long term perspective for sustainable development requires the broad based participation of various stakeholders in policy formulation, decision-making and implementation at all levels in particular of issues of biological diversity and this must be encouraged. While progress towards sustainable development has been made through meetings, agreements and changes in environmental governance, real change has been slow. To effectively address environmental problems, policy-makers should design policies that tackle both pressures and the drivers behind them. Economic instruments such as market creation and charge systems may be used to help spur environmentally sustainable behaviour.

Notes & References

1. William Ruckelshaus, Business Week, 18 June 1990.
2. Dr. Arundhati Kulkarni, "Biodiversity and Sustainable Development: A Critical Analysis" *International Journal of Scientific & Engineering Research*, Volume 3, Issue 4, April-2012, p.1.
3. Arvind Jasrotia, Environmental Protection and Sustainable Development: Exploring the Dynamics of Ethics and Law, (2007) *Journal of the Indian Law Institute*, Vol.49, p.34.
4. S.R. Myneni, Law of Intellectual Property', 5th Edn, (Asia Law House, Hyderabad, 2009), p.536.
5. Papers.ssrn.com/sol3/papers.cfm? abstractid=355141
6. V.R. Krishna Iyer, "The Dialectics and Dynamics of Human Rights in India", 1999, p.7.
7. www.suite101.com/content/the-importance- of - diversity-a214198
8. Business Leaders Endorse Sustainable Development Goals as Framework for Shaping Corporate Strategies (2016). Available at: https://www.unglobalcompact.org/news/3571-06-23-2016
9. *Printsipyi ustoychivogo razvitiya v deyatelnosti finansovyih institutov razvitiya i mezhdunarodnyih organizatsiy* [The principles of sustainable development in financial development institutions and international organizations] Quarterly Bulletin VEB, 2016. Vol.12, No.3, 17 p. (in Russian).
10. John C. Dernbach 1,* and Joel A. Mintz 2,Environmental Laws and Sustainability: An Introduction, (2011) pp.1-2, www.mdpi.com/journal/sustainability.
11. Ezeabasili, Nkechi (Legal Mechanism for Achieving Environmental Sustainability in Nigeria (pp. 369-370, IAARR, 2009.
12. No person shall be deprived of his life or personal liberty except according to the procedure established by law.
13. S.2 (a) of the Environment (Protection) Act, 1986.
14. Indian Council for Enviro Legal Action v. Union of India, (1996) 5 S.C.C. 281.
15. 15. A.IR. 1996 SC 2715.
16. Pratibha Singh, Anoop Singh, Piyush Malaviya ,*Text Book of Environment and Ecology*, ACME Learning Pvt. Ltd. ,1stEdition, 2009, p.50.
17. http://www.e-ir.info/2011/07/27/the-concept-of-sustainable-development/ visited on 24-12-2016 at 12.35 pm.
18. http://www.un-documents.net/our-common-future.pdf accessed on 25-12-2016 at 12.50 pm.

19. http://www.environment.gov.au/about-us/international/rio-20 visited on 25-12- 2016 at 1.10 pm
20. (1996) 5 SCC 647.
21. Kinkri Devi v. State of HP, AIR. 1988 H.P. 4.
22. AIR 1996 SC 149.
23. Ibid.
24. Principle 1-Man has the fundamental right to freedom, equality and adequate conditions of life, in an environment of a quality that permits a life of dignity and well-being, and he bears a solemn responsibility to protect and improve the environment for present and future generations.
25. Principle 2-the the natural resources of the earth, including the air, water, lands, flora and fauna and especially representative samples of natural ecosystems, must be safeguarded for the benefit of the present and future generations through careful planning or management , as appropriate.
26. AIR 1988 H.P.4.
27. (2013) 8 SCC 418.
28. Suo Motu Writ Petition (civil) No. 284 of 2012, (2013) 1 MLJ 639 (SC).
29. Principle 8 of Rio declaration.
30. AIR. 2002 Mad.298.
31. Dr. Paramjit S. Jaswal, Dr. Nishtha Jaswal, Vibhuti Jaswal, *Environmental Law*, Allahabad Law Agency, 4thEdn., 2015, p.139.
32. Ibid.
33. Principle 15 of Rio Declaration.
34. AIR 1999 SC 812.
35. The discharge of toxic substance or of other substance and the release of heat, in such quantities or concentration as to exceed the capacity of environment to render them harmless, must be halted in order to ensure that serious or irreversible damage is not inflicted upon ecosystems. The just struggle of the peoples of all countries against should be supported.
36. (1997) 2 SCC 353.
37. (2005) 13 SCC 156.
38. A.I.R. 1996 S.C.2715.
39. Dr. Paramjit S. Jaswal, Dr. Nishtha Jaswal, Vibhuti Jaswal, *Environmental Law*, Allahabad Law Agency, 4th Edn., 2015.
40. (1996) 3 SCC 212.
41. (2005) 13 SCC 156.

42. (2013) 4 SCC 575.
43. The states should co-operate to strengthen indigenous capacity-building for sustainable development by improving scientific understanding through exchange of scientific and technological knowledge, and by enhancing the development, adaptation, diffusion and transfer of technologies, including new and innovative technologies.
44. Environmental issues are best handled with the participation of all concerned citizens, at the relevant level.
45. The states should co-operate to promote a supportive and open international economic system that would lead to economic growth and sustainable development in all countries, to better address the problems of environmental degradation.
46. The states should co-operate to promote a supportive and open international economic system that would lead to economic growth and sustainable development in all countries, to better address the problems of environmental degradation.
47. http://www.un-documents.net/our-common-future.pdf, visited on 30-7-2018 at 10.15 am
48. The word socialist was added to the preamble by The Constitution (Forty Second) Amendment Act, 1976.
49. 1986 (2) CLJ 26.
50. Article 48-A.
51. AIR. 1988 Raj. 2.
52. Article 51 A (g).
53. AIR. 1987 A.P. 171.
54. (2002) 4 SCC 356.
55. (2004) 9 SCC 362.
56. AIR 1985 SC 652.
57. (2009) 17 SCC 764.

Books Referred

Arjya B. Majumdar, Debosmitha Nandi , Swayambhu Mukherjee , *Environment and Wildlife Laws in India*, Lexis Nexis,1st Edn.,2013.

Pratibha Singh, Anoop Singh, Piyush Malaviya, *A text Book of environment and Ecology*, Acme Learning, 1st Edn., 2009.

Dr. Paramjith S. Jasswal, Dr. Nishitha Jaswal,Vibhuti Jaswal, *Environmental law*, Allahabad Law Agency,4th Edn.,2015.

Indrajith Dube, *Environmental Jurisprudence*, Lexis Nexis Butterwoths, 2007.

Syam Divan, Armin Rosencranz, *Environmental Law and Policy in India*, Oxford University Press, 2nd Edn.2012.

P. Leelakrishnan, *Environmental Law in India*, Lexis Nexis, 3rd Edn., 2008.

Dr. N. Maheswara Swamy, *A Text Book on Environmental Law*, Asia Law House, 2nd Edn., 2008.

List of Abbreviations

A.I.R.: All India Reporter

A.P.: Andhra Pradesh

C.L.J.: Calcutta Law journal

H.C.: High Court

I.U.C.N.: International Union for Conservation of Nature

PIL: Public Interest Litigation

P.P.P.: Polluter Pays Principle

Raj. : Rajasthan

S.C.: Supreme Court

S.C.C.: Supreme Court Cases

U.N.: United Nations

U.N.C.H.E.: United Nations Commission on Human Environment

W.W.F.: World Wide Fund for Nature

Index

F

G

H

I

L

M

N

www.ingramcontent.com/pod-product-compliance
Ingram Content Group UK Ltd.
Pitfield, Milton Keynes, MK11 3LW, UK
UKHW021950270726
14060UKWH00002B/451